Lambert M. Surhone, Miriam T. Timpledon,
Susan F. Marseken (Ed.)

Vocaloid

Lambert M. Surhone, Miriam T. Timpledon,
Susan F. Marseken (Ed.)

Vocaloid

Yamaha Corporation, Lyrics, Melody, Miriam Stockley, NAMM Show, Computer Music

Betascript Publishing

Imprint

Permission is granted to copy, distribute and/or modify this document under the terms of the GNU Free Documentation License, Version 1.2 or any later version published by the Free Software Foundation; with no Invariant Sections, with the Front-Cover Texts, and with the Back-Cover Texts. A copy of the license is included in the section entitled "GNU Free Documentation License".

All parts of this book are extracted from Wikipedia, the free encyclopedia (www.wikipedia.org).

You can get detailed informations about the authors of this collection of articles at the end of this book. The editors (Ed.) of this book are no authors. They have not modified or extended the original texts.

Pictures published in this book can be under different licences than the GNU Free Documentation License. You can get detailed informations about the authors and licences of pictures at the end of this book.

The content of this book was generated collaboratively by volunteers. Please be advised that nothing found here has necessarily been reviewed by people with the expertise required to provide you with complete, accurate or reliable information. Some information in this book maybe misleading or wrong. The Publisher does not guarantee the validity of the information found here. If you need specific advice (f.e. in fields of medical, legal, financial, or risk management questions) please contact a professional who is licensed or knowledgeable in that area.

Any brand names and product names mentioned in this book are subject to trademark, brand or patent protection and are trademarks or registered trademarks of their respective holders. The use of brand names, product names, common names, trade names, product descriptions etc. even without a particular marking in this works is in no way to be construed to mean that such names may be regarded as unrestricted in respect of trademark and brand protection legislation and could thus be used by anyone.

Cover image: www.PureStockX.com
Concerning the licence of the cover image please contact PureStockX.

Contact:
VDM Publishing House Ltd.,17 Rue Meldrum, Beau Bassin,1713-01 Mauritius
Email: info@vdm-publishing-house.com
Website: www.vdm-publishing-house.com

Published in 2010

Printed in: U.S.A., U.K., Germany. This book was not produced in Mauritius.

ISBN: 978-613-0-53340-3

Contents

Articles

References

Article Licenses

Vocaloid

Developer(s)	Yamaha Corporation
Initial release	January 2004
Stable release	Vocaloid 2
Operating system	Windows XP / Vista
Available in	Japanese, English
Development status	Active
Type	Musical Synthesizer Application
License	proprietary
Website	[1]

Vocaloid is a singing synthesizer application software developed by the Yamaha Corporation that enables users to synthesize singing by typing in lyrics and melody.

Development history

Yamaha announced its development in 2003 and on January 15, 2004, Leon and Lola, the first Vocaloid products were launched. They were not released as Yamaha products, but as Vocaloid Singer Libraries, developed by third party developers, the products were powered by the Vocaloid software, under license from Yamaha. Leon, Lola, and Miriam (Miriam using the voice of Miriam Stockley) have been released from Zero-G Limited,[2] UK, while Meiko (released on October 5, 2004 and using vocal samples from the Japanese singer Meiko Haigo[3]) and Kaito (released on February 17, 2006 and sampled from Naoto Fuuga) have been released from Crypton Future Media, Japan.[4] [5]

In January 2007, Yamaha announced a new version of the software engine, Vocaloid2, with various major improvements in usability and synthesis quality. Zero-G and others announced products powered by the new software engine in early 2007. PowerFX released the first Vocaloid2 package in June 2007, an English product named Sweet Ann. This was shortly followed in August 2007, when Crypton released Hatsune Miku, the first in a series of Japanese Vocaloid 2 character voices. The second package Kagamine Rin/Len was released on December 27, 2007 and the updated edition "act2" was released in July 2008. The first Vocaloid 2 product from Zero-G, Vocaloid Prima, an English classical voice, was finally released on January 14, 2008 in the UK[6] and February 22, 2008 in Japan. It was originally scheduled for release in spring 2007. Prima was introduced at the NAMM Show 2008;[7] . The third Vocaloid2 product from Crypton, Megurine Luka, went on sale on January 30, 2009. She is the first bilingual Vocaloid product, capable of singing in both Japanese and English.

Products based on Vocaloid

Vocaloid

- Leon: English male (March 3, 2004)
- Lola: English female (March 3, 2004)
- Miriam: English female (July 26, 2004)
- Meiko: Japanese female (November 5, 2004)
- Kaito: Japanese male (February 17, 2006)

Vocaloid 2

- Character Vocal Series
  - Hatsune Miku: Japanese female (August 31, 2007)
  - Kagamine Rin/Len: Japanese female and male respectively (December 27, 2007)
  - Megurine Luka: Japanese and English female (January 30, 2009)
- Gackpoid: Japanese male (July 31, 2008)
- Megpoid: Japanese female (June 25, 2009)
- Sweet Ann: English female (June 29, 2007)
- Prima: English female (January 14, 2008)
- Big-Al: English male (December 22, 2009)
- Sonika: English Female (July 14, 2009)[8]
- SF-A2 Miki: Japanese female (December 4, 2009)[9]
- Kaai Yuki: Japanese female (December 4, 2009)[10]
- Hiyama Kiyoteru: Japanese male (December 4, 2009)[11]
- Tonio: English male

Crypton Future Media's Character Vocal Series

The Character Vocal Series is a computer music program that synthesizes singing in Japanese. Developed by Crypton Future Media, it utilizes Yamaha's Vocaloid2 technology with specially recorded vocals of voice actors. To create a song, the user must input the melody and lyrics. A piano roll type interface is used to input the melody and the lyrics can be entered on each note. The software can change the stress of the pronunciations, add effects such as vibrato, or change the dynamics and tone of the voice.

The series is intended for professional musicians as well as light computer music users. The programmed vocals are designed to sound like an idol singer from the future. According to Crypton, because professional singers refused to provide singing data, in fear that the software might create their singing voice's clones, Crypton changed their focus from imitating certain singers to creating characteristic vocals. This change of focus led to sampling vocals of voice actors.[12]

Each Japanese Vocaloid is given an anime-type character with specifications on age, height, weight, and musical strengths (genre, pitch range and ideal tempos). The characters of the first three installments of the series are created by illustrator Kei.

Any rights or obligations arising from the vocals created by the software belong to the software user. Just like any music synthesizer, the software is treated as a musical instrument and the vocals as sound. Under the term of license, the Character Vocal Series software can be used to create vocals for commercial or non-commercial use as long as the vocals do not offend public policy. In other words, the user is bound under the term of license with Crypton not to synthesize derogatory or disturbing lyrics. On the other hand, copyrights to the mascot image and name belong to Crypton. Under the term of license, a user cannot commercially distribute a vocal as a song sung by the character, nor use the mascot image on commercial products, without Crypton's consent.

Hatsune Miku

Hatsune Miku (初音ミク) is the first installment in the Vocaloid2 Character Vocal Series released on August 31, 2007. The name of the title and the character of the software was chosen by combining **Hatsu** (初 *First*), **Ne** (音 *Sound*), and **Miku** (未来 *Future*).[13] The data for the voice was created by actually sampling the voice of Saki Fujita, a Japanese voice actress. Unlike general purpose speech synthesizers, the software is tuned to create J-pop songs commonly heard in anime, but it is possible to create songs from other genres.

The cover of the first release.

Nico Nico Douga played a fundamental role in the recognition and popularity of the software. Soon after the release of the software, users of Nico Nico Douga started posting videos with songs created by the software. According to Crypton, a popular video with a comically-altered Miku holding a leek, singing Ievan Polkka, presented multifarious possibilities of applying the software in multimedia content creation.[14] As the recognition and popularity of the software grew, Nico Nico Douga became a place for collaborative content creation. Popular original songs written by a user would generate illustrations, animation in 2D and 3D, and remixes by other users. Other creators would show their unfinished work and ask for ideas.[15]

On October 18, 2007, an Internet BBS website reported Hatsune Miku was suspected to be victim of censorship by Google and Yahoo!, since images of Miku did not show up on the image searches.[16] Google and Yahoo denied any censorship on their part, blaming the missing images on a bug that does not only affect "Hatsune Miku" but other search keywords as well. Both companies expressed a willingness to fix the problem as soon as possible.[17] Images of Miku were relisted on Yahoo on October 19, 2007.

A Hatsune Miku manga called *Maker Hikōshiki Hatsune Mix* began serialization in the Japanese manga magazine *Comic Rush* on November 26, 2007, published by Jive. The manga is drawn by Kei, the original character designer for Hatsune Miku. A second manga called *Hachune Miku no Nichijō Roipara!* drawn by Ontama began serialization in the manga magazine *Comp Ace* on December 26, 2007, published by Kadokawa Shoten.

The character's first appearance in an anime is in *(Zoku) Sayonara Zetsubō Sensei*, where she (and various other people and characters) try out to be the voice of Meru Otonashi. For online multi-player games, the Japanese version of *PangYa* started a campaign with Hatsune Miku on May 22, 2008 in which a player could purchase her outfit for one of the characters.[18] [19] . Her first appearance in a video game is in *13-sai no Hello Work DS* (13歳のハローワークDS) for the Nintendo DS where she is included as one of the characters.[20] [21] Hatsune Miku was given a PlayStation Portable game called *Hatsune Miku: Project DIVA* that was released on July 2, 2009 by Sega.[22] Hatsune Miku made a cameo appearance in the *Lucky Star* OVA in the form of Kagami's cosplay in her dream. She made a vocal appearance in the finale ending theme of the anime *Akikan!*. Hatsune Miku received the 2008 Seiun Award in the free category.[23] [24] She is also a playable character in the game *Tales of Graces* as downloadable content for 400 Wii points.[25]

On August 27, 2008, Victor Entertainment released the album *Re:package* which contains a collection of songs performed by Hatsune Miku and composed by a pair of dōjin artists named Livetune. The album sold over 20,000 copies in its first week and successfully broke into Oricon's charts by placing fifth for the week.[26] Following up with the success of *Re:package*, Victor Entertainment released Livetune's second Hatsune Miku album, *Re:MIKUS*, on March 25, 2009, which contains many remixed versions of original songs from various original music artists, such as Supercell and kz of Livetune.[27] It also contains four more original songs sung by Hatsune Miku, which again were made by original dōjin artists.

As a virtual idol, Hatsune Miku performed a "live" concert during Animelo Summer Live on August 23, 2009[28] and at Anime Festival Asia (AFA), Singapore in 2009.[29] In addition, singer Gackt performed alongside Miku.

In late November 2009, a petition was launched in order to get a custom made Hatsune Miku aluminum plate (8 cm x 12 cm, 3.1" x 4.7") made that will be used as a balancing weight for the Japanese Venus spacecraft explorer Akatsuki which will be launched in summer 2010.[30] Started by Hatsune Miku fan Sumio Morioka that goes by chodenzi-P, this project has received the backing of Dr. Seiichi Sakamoto of the Japan Aerospace Exploration Agency. On December 22, 2009, the petition exceeded the needed 10,000 signatures necessary to have the plates made. An original deadline of December 20, 2009 had been set to send in the petition, but due to a couple of delays in the Akatsuki project, a new deadline of January 6, 2010 was set; by this deadline, over 14,000 signatures had been received.

Kagamine Rin and Len

Released on December 27, 2007, **Kagamine Rin/Len** (鏡音リン・レン) is the second installment of the Vocaloid2 Character Vocal Series. Their surname was chosen by combining **Kagami** (鏡 *Mirror*), **Ne** (音 *Sound*), with the first syllables of their given names a pun on "Left" and "Right". According to Vocaloid's official blog, the package includes two voice banks: one for Rin and another for Len, both provided by the voice actor Asami Shimoda. Despite the double voice banks, the package still sells at the same price as Hatsune Miku.[31] Their only cameo appearance in an anime is in *(Zoku) Sayonara Zetsubō Sensei*, where the two, Miku, Kaito, and Meiko (and various other people and characters) try out to be the voice of Meru Otonashi.

On June 12, 2008, Crypton announced the updated edition, named "act2", will be released in early July 2008. Users who had bought the old version will get an expansion disc free of charge. On June 18, 2008, beta demonstration songs using the new version were released on the company's official blog.[32] The expansion disc is an entirely different software and does not affect the original Kagamine Rin/Len installation in any way, giving the user options to either use the old or new voice sets exclusively or combine their usage.

Megurine Luka

The third installment in the character vocal series, **Megurine Luka** (巡音ルカ), was released on January 30, 2009.[33] Her surname combines **Meguri** (巡 *Circulate*) and **Ne** (音 *Sound*). Luka's voice is that of a twenty-year-old female and she can sing in both Japanese and English. Her voice bank was sampled from Yū Asakawa. The manga artist Kei, who illustrated Miku, Rin, and Len, also designed her mascot. However, unlike previous mascots in the series, her costume is not based on a school uniform but more based on a modern style cheongsam.

Involvement in Super GT Series

In the 2008 season, two cars competing in the GT300 class adopted images and color schemes involving characters in the Vocaloid series. The first car, named "Hatsune Miku Studie Glad BMW Z4" (#808), was used by Studie (a tuning shop for BMW).[34] It used Hatsune Miku's image and color scheme, and has its debut in round six in the Suzuka Circuit. It marked the first so-called "Itasha" to participate in international class races under FIA. Though it never qualified in the qualify session mainly due to fuel problems that season (including the miss of its Suzuka debut due to a misunderstanding of the rules), it was

The Hatsune Miku Studie Glad BMW Z4 competed in the 2008 Super GT season.

allowed to race in the final round at Mt. Fuji, and completed the race in eighteenth place. Crypton fully supported the process of participation. The car attracted many motor sport and non-motor sport fans not only because of its color scheme, but also marked as a testing point of BMW's return in the Super GT series, since Z4 has already dominated in the Super Taikyu series in Japan. After the final race at Mt. Fuji, it is known that Studie will continue to adopt Hatsune Miku's image in the following season.[35]

In the final round at Mt. Fuji, one of the leading teams, Mola, adopted images of Kagamine Rin and Len in their "Mola Leopalace Z" (#46); they finished sixth in the race. Unlike the previous car, it did not change its name in the process.[36] [37]

On top of that, Studie was one of the few teams that adopted color designs from the general public, rather than a professional designer in international motorsport history. When Piapro (official fansite which is in charge of collecting designs) was collecting designs of the Hatsune Miku Studie Glad BMW Z4, they did not announce the designs would be used in Super GT series, instead most of the designers expected it would be racing in the lower-level Super Taikyu Series, or as display cars in autoshows or Comiket (the plan was disguised as a model-car design contest), so many of them chose #39, the number which usually belongs to Hatsune Miku. However, in the Super GT series, the #39 belonged to Toyota Team SARD (Now known as Lexus team SARD) in the GT500 class; this fact was reflected in Goodsmile Racing stickers for Z4 car models released after the 2008 season, which they provided both for #39 and #808. Like the previous season, Piapro picked a design from the public and it was revealed in February 2009.[38]

Internet Co. Ltd.'s Vocaloids

Gackpoid

Internet Co. Ltd. wanted to utilize the voice of a musician for the creation of Vocaloid but felt it would be difficult to acquire cooperation. They consulted Dwango, who suggested Gackt, a musician and an actor, as he had previously provided his voice for Dwango's cell phone services.[39] He lent his voice and named the Vocaloid, **Gackpoid** (がくっぽいど Gakuppoido). The product was originally intended to be released in June 2008, but although Gackt existed as a model for the Vocaloid, its illustrated avatar was yet to be determined. Finally a popular manga author Kentarō Miura, famous for his dark fantasy epic *Berserk*, was asked for his cooperation. Due to Miura's affection for Nico Nico Douga, he agreed to offer his services as a character designer for free. As a fan of *Berserk*, Gackt was more than happy with this arrangement, and requested Miura's sketches be faxed to him as well as the developers, even though he was on location for the filming of Guy Moshe's *Bunraku*.[40] Gackpoid was released on July 31, 2008.[41] Gackpoid includes a new program, OPUS Express, for mixing vocal parts with accompaniment or phoneme data.[42] Two of Gackt's songs and other three songs are also included as samples.[41] Miura's design for Gackpoid was named **Kamui Gakupo/Gackpo** (神威がくぽ) after the stage name of Gackt and has a samurai aesthetic—the character is clad in Jinbaori, a kind of kimono which was used as a battle surcoat, and carries a katana that somehow acts as a musical instrument.[43]

Megpoid

Internet Co. published their second Vocaloid software titled **Megpoid** (メグッポイド) on June 26, 2009 using the voice of Megumi Nakajima. She has bright green hair and wears red goggles on her head. This is a parody of both her theme color and a character that Nakajima voiced, Ranka Lee from *Macross Frontier*; her design is similar to Ranka Lee.[44] One of her demo songs is "Be Myself", an original song by Nakajima. Megpoid sample files are included in the disc for the software.Her voice range is F2-A4 and her optimum tempo is 60-175BPM. Its character was named **GUMI** (ぐみ), which was designed by the manga artist Yuki Masami.

External links

- Official website [45]
- Crypton's official Vocaloid2 website [46] **(Japanese)**
- PowerFX official website [47]

References

[1] http://www.vocaloid.com/index.en.html
[2] "NEWS: ZERO-G announces first 3 Vocaloid titles" (http://www.zero-g.co.uk/index.cfm?articleid=803). . Retrieved 2008-07-17.
[3] "Vocaloid's English official website" (http://www.vocaloid.com/en/index.html). Yamaha. 2004-11-09. . Retrieved 2008-09-01.
[4] "Crypton Future Media Vocaloid MEIKO" (http://www.kvraudio.com/get/1368.html). . Retrieved 2008-07-17.
[5] "Crypton Vocaloid KAITO" (http://www.crypton.co.jp/mp/do/prod?id=27720) (in Japanese). Crypton. . Retrieved 2008-07-17.
[6] "Zero-G shipping Vocaloid PRIMA" (http://rekkerd.org/zero-g-shipping-vocaloid-prima/). 2008-01-14. . Retrieved 2008-09-01.
[7] "EASTWEST Introduces Zero-G Vocaloid Prima At NAMM 2008" (http://namm.harmony-central.com/WNAMM08/Content/EastWest/
PR/Vocaloid-Prima.html). 2008-01-23. . Retrieved 2008-09-01.
[8] "Zero-G shipping Vocaloid Sonika" (http://rekkerd.org/zero-g-shipping-vocaloid-sonika/). Rekkerd. 2009-07-15. . Retrieved 2009-07-23.
[9] "SF-A2 開発コード miki [SF-A2 Development Code miki]" (http://www.ah-soft.com/vocaloid/miki/index.html) (in Japanese).
AH-Software. . Retrieved 2009-11-18.
[10] "ボカロ小学生 歌愛ユキ [Voiceroid Elementary School Student Kaai Yuki]" (http://www.ah-soft.com/vocaloid/yuki/index.html) (in
Japanese). AH-Software. . Retrieved 2009-11-18.
[11] "ボカロ先生 氷山キヨテル [Voiceroid Teacher Hiyama Kiyoteru]" (http://www.ah-soft.com/vocaloid/kiyoteru/index.html) (in
Japanese). AH-Software. . Retrieved 2009-11-18.
[12] "How Hatsune Miku was born: Interview with Crypton Future Media" (http://www.itmedia.co.jp/news/articles/0802/22/news013.
html) (in Japanese). 2008-02-22. . Retrieved 2008-02-28.
[13] "Exceptional sales of Hatsune Miku" (http://www.itmedia.co.jp/news/articles/0709/12/news035.html) (in Japanese). 2007-11-12. .
Retrieved 2007-11-08.
[14] "How Hatsune Miku opened the creative mind: Interview with Crypton Future Media" (http://www.itmedia.co.jp/news/articles/0802/
25/news017.html) (in Japanese). 2008-02-25. . Retrieved 2008-02-29.
[15] "DTM in the boom again: How anonymous creators are discovered by Hatsune Miku" (http://www.itmedia.co.jp/news/articles/0709/
28/news066.html) (in Japanese). 2007-09-28. . Retrieved 2008-02-29.
[16] "Hatsune Miku images disappearing from the Internet" (http://www.itmedia.co.jp/news/articles/0710/18/news040.html) (in Japanese).
2007-10-18. . Retrieved 2007-10-21.
[17] "Google and Yahoo "investigating the problem" on the disappearance of Hatsune Miku" (http://www.itmedia.co.jp/news/articles/0710/
18/news065.html) (in Japanese). 2007-10-18. . Retrieved 2007-11-23.
[18] "Announcement of *PangYa* season 4" (http://www.onlinegamer.jp/news/4750/) (in Japanese). OnlineGamer. 2008-05-05. . Retrieved
2008-06-06.
[19] "*PangYa* cooperation with Hatsune Miku" (http://www.famitsu.com/pcent/news/1215366_1341.html) (in Japanese). Famitsu.
2008-05-22. . Retrieved 2008-06-06.
[20] "Hatsune Miku appears in *13-sai no Hello Work DS*" (http://www.famitsu.com/game/coming/1213533_1407.html) (in Japanese).
Famitsu. 2008-02-15. . Retrieved 2008-06-06.
[21] "Composing music with Hatsune Miku in *13-sai no Hello Work DS*" (http://www.famitsu.com/game/coming/1213865_1407.html) (in
Japanese). Famitsu. 2008-03-03. . Retrieved 2008-06-06.
[22] *Hatsune Miku: Project Diva* official website" (http://miku.sega.jp/) (in Japanese). Sega. . Retrieved 2009-03-15.
[23] "Seiun Award: Long Novel Goes to *Toshokan Sensō*, and Hatsune Miku, *20th Century Boys*, and Others Win" (http://mainichi.jp/enta/
mantan/news/20080825mog00m200023000c.html) (in Japanese). Mainichi Shimbun. 2008-08-25. . Retrieved 2008-08-31.
[24] *Library War, Dennō Coil, 20th Century Boys* Win Seiun Awards" (http://www.animenewsnetwork.com/news/2008-08-24/
library-war-denno-coil-20th-century-boys-win-seiun-awards). Anime News Network. 2008-08-24. . Retrieved 2008-08-31.
[25] "First Look: Tales of Graces Hatsune Miku DLC" (http://www.andriasang.com/e/blog/2009/12/18/
tales_of_graces_hatsune_miku_dlc). Andriasang. December 18, 2009. . Retrieved December 18, 2009.
[26] "Album featuring "Hatsune Miku" Top 10" (http://www.oricon.co.jp/news/rankmusic/57783/) (in Japanese). Oricon. 2008-09-02. .
Retrieved 2008-09-13.
[27] "Livetune feat. Hatsune Miku special site" (http://www.jvcmusic.co.jp/livetune_feat_hatsunemiku/) (in Japanese). . Retrieved
2009-06-06.
[28] "Hatsune Miku Virtual Idol Performs 'Live' Before 25,000" (http://www.animenewsnetwork.com/news/2009-08-23/
hatsune-miku-virtual-idol-performs-live-before-25000). Anime News Network. 2009-08-23. . Retrieved 2009-08-26.
[29] "Virtual Idol "Hatsune Miku" to perform overseas at "I LOVE anisong" stage!" (http://afa09.com/i_love_miku.html). Anime Festival
Asia. . Retrieved 2009-11-24.

[30] "金星探査機「あかつき」に初音ミクΩを搭載する署名 [Sign to Get Hatsune Miku Image on Board Venus Explorer Akatsuki]" (https://spreadsheets.google.com/viewform?hl=en&formkey=dEhaZ1JMUXI3MEl4Qk14VXBCZXVHdlE6MA) (in Japanese). Google. . Retrieved December 18, 2009.

[31] "Vocaloid2 info: CV02 "Kagamine Rin/Len" announced" (http://blog.crypton.co.jp/mp/2007/12/vocaloid2_cv02_3.html) (in Japanese). 2007-12-03. . Retrieved 2007-12-04.

[32] "Rin/Len act2 beta demonstration songs released" (http://blog.crypton.co.jp/mp/2008/06/vocaloid2_act2_1.html) (in Japanese). Crypton. 2008-06-18. . Retrieved 2008-06-24.

[33] "Megurine Luka Announced as Next Vocaloid 2 Character" (http://www.animenewsnetwork.com/news/2009-01-06/megurine-luka-announced-as-next-vocaloid-2-character). Anime News Network. 2009-01-06. . Retrieved 2009-01-07.

[34] "The Rumored No. 808 Hatsune Miku Studie Glad BMW Z4 Latest News" (http://ww2.supergt.net/gtcgi/prg/NList02.dll/Code?No=NS010942&List=13) (in Japanese). Super GT.net. 2008-08-18. . Retrieved 2008-08-21.

[35] "Itasha storm in Super GT. Report of Final Race at Mt. Fuji (Page 3)" (http://ascii.jp/elem/000/000/187/187441/index-3.html) (in Japanese). ASCII. 2008-11-11. . Retrieved 2008-11-16.

[36] "Itasha storm in Super GT. Report of Final Race at Mt. Fuji (page 2)" (http://ascii.jp/elem/000/000/187/187441/index-2.html) (in Japanese). ASCII. 2008-11-11. . Retrieved 2008-11-16.

[37] "2008 Super GT Champions" (http://supergt.net/supergt/2008/08series/index_e.htm). Super GT. 2008-11-11. . Retrieved 2008-11-16.

[38] "Hatsune Miku BMW Z4 comes back stronger to GT 300 for 2009 season!" (http://ww2.supergt.net/gtcgi/prg/NList02.dll/Code?No=NS012104&List=13). Super GT. 2008-02-16. . Retrieved 2008-02-18.

[39] Michiko Nagai (2008-06-20). "Gackt to Sing and Kentarō Miura to Draw Gackpoid" (http://japan.cnet.com/news/tech/story/0,2000056025,20376132,00.htm) (in Japanese). CNET Japan. . Retrieved 2008-06-29.

[40] "Gackpoid to be Sold in Late July; Kentarō Miura to Illustrate" (http://www.itmedia.co.jp/news/articles/0806/20/news043.html) (in Japanese). ITMedia. 2008-06-20. . Retrieved 2008-06-29.

[41] "Gackpoid" (http://www.ssw.co.jp/products/vocal/gackpoid/index.html) (in Japanese). Internet Co. Ltd.. 2008-06-20. . Retrieved 2008-06-29.

[42] "Kamui Gakupo, Debut At the End of July! "Gackpoid"" (http://www.barks.jp/news/?id=1000041036) (in Japanese). Barks. 2008-06-20. . Retrieved 2008-06-29.

[43] "gackpoid-official"EpisodeII Character Design"" (http://www.ssw.co.jp/products/vocal/gackpoid/infomation/episode2.html) (in Japanese). Internet Co. Ltd.. 2008-07-31. . Retrieved 2009-11-29.

[44] "Megpoid official listing" (http://www.ssw.co.jp/products/vocal/megpoid/index.html) (in Japanese). Internet Co.. . Retrieved 2009-05-14.

[45] http://www.vocaloid.com/en/index.html
[46] http://www.crypton.co.jp/mp/pages/prod/vocaloid/
[47] http://www.powerfx.com/

Yamaha Corporation

	YAMAHA	
Type	Public (TYO: 7951 [1])	
Founded	October 12, 1887	
Headquarters	Hamamatsu, Shizuoka prefecture, Japan	
Industry	Conglomerate	
Products	Musical instruments, Audio/Video, Electronics, Computer related products, ATVs, Motorbikes, Vehicle Engines, Personal water craft, golf clubs	
Revenue	▯ 4.676 billion US$ (March 31, 2009)[2]	
Operating income	140.95 million US$ (March 31, 2009)	
Net income	▯(209.87) million US$ (March 31, 2009)	
Employees	26,803 (March 31, 2009)	
Website	Yamaha.com [3]	

The **Yamaha Corporation** (ヤマハ株式会社 *Yamaha Kabushiki Gaisha*) (TYO: 7951 [1]) is a multinational corporation and conglomerate based in Japan with a wide range of products and services, predominantly musical instruments, motorcycles, power sports equipment, and electronics.

History

Yamaha was established in 1887 as a piano and reed organ manufacturer by Torakusu Yamaha as **Nippon Gakki Company, Limited** (日本▯器製造株式会社 *Nippon Gakki Seizō Kabushiki Gaisha*) (literally Japan Musical Instrument Manufacturing Co.) in Hamamatsu, Shizuoka prefecture, and was incorporated on October 12, 1897. The company's origins as a musical instrument manufacturer is still reflected today in the group's logo—a trio of interlocking tuning forks.[4]

The headquarters of Yamaha Corporation

After World War II, company president Genichi Kawakamisaki repurposed the remains of the company's war-time production machinery and the company's expertise in metallurgical technologies to the manufacture of motorcycles. The YA-1 (AKA Akatombo, the "Red Dragonfly"), of which 125 were built in the first year of production (1958), was named in honor of the founder. It was a 125cc, single cylinder, two-stroke, street bike patterned after the German DKW RT125 (which the British munitions firm, BSA, had also copied in the post-war era and manufactured as the *Bantam* and Harley-Davidson as the *Hummer*). In 1959, the success of the YA-1 resulted in the founding of the Yamaha Motor Co., Ltd.

Yamaha has grown to become the world's largest manufacturer of musical instruments (including pianos, "silent" pianos, drums, guitars, brass instruments, woodwinds, violins, violas, celli, vibraphones, and saxophones), as well as a leading manufacturer of semiconductors, audio/visual, computer related products, sporting goods, home appliances, specialty metals, and industrial robots.

In October 1987, on its 100th anniversary, the name was changed to *The Yamaha Corporation*.

In 1989, Yamaha shipped the world's first CD recorder.

Yamaha purchased Sequential Circuits in 1988 and bought a significant share (51%) of competitor Korg in 1989–1993.

In 2002, Yamaha closed down its archery product business that was started in 1959. Six archers in five different Olympic Games won gold medals using their products.[5]

It acquired German audio software manufacturers Steinberg in 2004, from Pinnacle Systems.

In July, 2007, Yamaha bought out the minority shareholding of the Kemble family in Yamaha-Kemble Music (UK) Ltd, Yamaha's UK import and musical instrument and professional audio equipment sales arm, the company being renamed Yamaha Music U.K. Ltd in autumn 2007.[6] Kemble & Co. Ltd, the UK piano sales & manufacturing arm was unaffected.[7]

On December 20, 2007, Yamaha made an agreement with the Austrian Bank BAWAG P.S.K. Group BAWAG to purchase all the shares of Bösendorfer[8] , intended to take place in early 2008. Yamaha intends to continue manufacturing at the Bösendorfer facilities in Austria.[9] The acquisition of Bösendorfer was announced after the NAMM Show in Los Angeles, on January 28, 2008. As of February 1, 2008, Bösendorfer Klavierfabrik GmbH operates as a subsidiary of Yamaha Corp.[10]

Yamaha Corporation is also widely known for their music teaching programme that began in the 1980s.

Yamaha electronic pianos continue to be a successful, popular and respected product. For example the Yamaha YPG-625 was given the award "Keyboard of the Year" and "Product of the Year" in 2007 from *The Music and Sound Retailer* magazine [11].

Other companies in the Yamaha group include:

- Bösendorfer Klavierfabrik GmbH, Vienna, Austria.
- Yamaha Motor Company
- Yamaha Fine Technologies Co., Ltd.
- Yamaha Livingtec Corporation
- Yamaha Metanix Corporation
- Yamaha Pro Audio

Corporate mission

Kandō (感動) is a Japanese word, used by Yamaha to describe their corporate mission. Kandō in translation describes the sensation of profound excitement and gratification derived from experiencing supreme quality and performance.[12]

Sports teams

- Yamaha Jubilo—rugby

External links

- Official website [13]
- Yamaha Musician [14] - Yamaha enthusiast keyboard site
- Yamaha Piano [15]
- American artist program [16]

References

[1] http://www.bloomberg.com/apps/quote?ticker=7951:JP

[2] http://www.global.yamaha.com/ir/publications/pdf-data/2009/fin/fd-2009.pdf

[3] http://www.yamaha.com

[4] http://www.global.yamaha.com/about/brand/index.html

[5] "YAMAHA to Close Archery Products Business" (http://www.global.yamaha.com/news/2002/20020201.html). 2002-02-01.. Retrieved 2008-04-30.

[6] Cancellation of Joint Venture Contracts for Sales Subsidiaries in U.K. and Spain (http://www.global.yamaha.com/news/2007/20070710. html), Yamaha Global website, July 10, 2007

[7] Yamaha buys out Kemble family (http://www.mi-pro.co.uk/news/28011/Yamaha-buys-out-Kemble-family), MI Pro, July 10, 2007

[8] Competition For Bosendorfer. (http://www.forbes.com/2007/11/30/bosendorfer-bawag-yamaha-markets-equity-cx_jc_1130markets04. html)

[9] Yamaha Reaches Basic Agreement with Austrian Bank to Purchase All Shares of Bösendorfer (http://www.global.yamaha.com/news/ 2007/20071220a.html), Yamaha Global website, December 20, 2007

[10] Business Week. March 3, 2008. Bosendorfer Klavierfabrik GmbH. (http://investing.businessweek.com/businessweek/research/stocks/ private/snapshot.asp?privcapId=1554618)

[11] http://www.yamaha.com/yamahavgn/CDA/ContentDetail/ModelSeriesDetail/ 0,,CNTID%25253D65229%252526CTID%25253D205200,00.html

[12] Yamaha Motor UK (http://www.yamaha-motor.co.uk/corporate/ymuk/mission.jsp) Yamaha Corporate Mission.

[13] http://www.yamaha.com/

[14] http://www.yamahamusician.com

[15] http://www.yamahapiyano.com/

[16] http://www.yamaha.com/yasi

Lyrics

Lyrics (in singular form **Lyric**) are a set of words that make up a song. The writer of lyrics is a lyricist or lyrist. The meaning of lyrics can either be explicit or implicit. Some lyrics are abstract, almost unintelligible, and, in such cases, their explication emphasizes form, articulation, meter, and symmetry of expression. The lyricist of traditional musical forms such as Opera is known as a librettist.

Etymology and usage

Lyric derives from the Greek word *lyrikos*, meaning "singing to the lyre".[1] A lyric poem is one that expresses a subjective, personal point of view.

The word *lyric* came to be used for the "words of a song"; this meaning was recorded in 1876.[1] The common plural (perhaps because of the association between the plurals *lyrics* and *words*), predominates contemporary usage. Use of the singular form *lyric* remains grammatically acceptable, yet remains considered erroneous in referring to a singular song word as a *lyric*.

Copyright and royalties

See Royalties

Currently, there are many websites featuring song lyrics. This offering, however, is controversial, since some sites include copyrighted lyrics offered without the holder's permission. The U.S. Music Publishers' Association (MPA), which represents sheet music companies, launched a legal campaign against such websites in December 2005, the MPA's president, Lauren Keiser, said the free lyrics web sites are "completely illegal" and wanted some website operators jailed.[2]

Academic study

- Lyrics can be studied from an academic perspective. For example, some lyrics can be considered a form of social commentary. Lyrics often contain political, social and economic themes as well as aesthetic elements, and so can connote messages which are culturally significant. These messages can either be explicit or implied through metaphor or symbolism. Lyrics can also be analyzed with respect to the sense of unity (or lack of unity) it has with its supporting music. Analysis based on tonality and contrast are particular examples.
- Chinese lyrics (詞) are Chinese poems written in the set metrical and tonal pattern of a particular song.

Riskiest Search

McAfee claims searches for phrases containing "lyrics" and "free" are the most likely to have risky results. [3]

See also

- Lyricist, a lyrics writer
- Instrumental, music without voice
- Libretto, the name used for the text of traditional music forms like opera

References

[1] Online Etymology Dictionary. Retrieved 2008-08-23 (http://www.etymonline.com/index.php?term=lyric)
[2] "Song sites face legal crackdown" (http://news.bbc.co.uk/1/hi/entertainment/4508158.stm) BBC News, 12 December 2005. Site accessed 7 January 2007
[3] http://us.mcafee.com/en-us/local/docs/most_dangerous_searchterm_us.pdf

Melody

A **melody** (from Greek μελῳδία - *melōidía*, "singing, chanting"), also **tune**, **voice**, or **line**, is a linear succession of musical tones which is perceived as a single entity. In its most literal sense, a melody is a sequence of pitches and durations, while, more figuratively, the term has occasionally been extended to include successions of other musical elements such as tone color.

Melodies often consist of one or more musical phrases or motifs, and are usually repeated throughout a song or piece in various forms. Melodies may also be described by their melodic motion or the pitches or the intervals between pitches (predominantly conjunct or disjunct or with further restrictions), pitch range, tension and release, continuity and coherence, cadence, and shape.

Elements

Given the many and varied elements and styles of melody "many extant explanations [of melody] confine us to specific stylistic models, and they are too exclusive."[1] Paul Narveson claimed in 1984 that more than three-quarters of melodic topics had not been explored thoroughly.[2]

The melodies existing in most European music written before the 20th century, and popular music throughout the 20th century, featured "fixed and easily discernible frequency patterns", recurring "events, often periodic, at all structural levels" and "recurrence of durations and patterns of durations".[1]

Melodies in the 20th century have "utilized a greater variety of pitch resources than has been the custom in any other historical period of Western music." While the diatonic scale is still used, the twelve-tone scale became "widely employed."[1] Composers also allotted a structural role to "the qualitative dimensions" that previously had been "almost exclusively reserved for pitch and rhythm". DeLone states, "The essential elements of any melody are duration, pitch, and quality (timbre), texture, and loudness.[1] Though the same melody may be recognizable when played with a wide variety of timbres and dynamics, the latter may still be an "element of linear ordering"[1]

Examples

Different musical styles use melody in different ways. For example:

"Pop Goes the Weasel" melody

- Jazz musicians use the melody line, called the "lead" or "head", as a starting point for improvisation.
- Rock music, melodic music, and other forms of popular music and folk music tend to pick one or two melodies (verse and chorus) and stick with them; much variety may occur in the phrasing and lyrics.
- Indian classical music relies heavily on melody and rhythm, and not so much on harmony as the above forms.
- Balinese gamelan music often uses complicated variations and alterations of a single melody played simultaneously, called heterophony.
- In western classical music, composers often introduce an initial melody, or theme, and then create variations. Classical music often has several melodic layers, called polyphony, such as those in a fugue, a type of counterpoint. Often, melodies are constructed from motifs or short melodic fragments, such as the opening of Beethoven's Fifth Symphony. Richard Wagner popularized the concept of a *leitmotif*: a motif or melody associated with a certain idea, person or place.
- While in both most popular music and classical music of the common practice period pitch and duration are of primary importance in melodies, the contemporary music of the 20th and 21st centuries pitch and duration have lessened in importance and quality has gained importance, often primary. Examples include musique concrete, klangfarbenmelodie, Elliott Carter's *Eight Etudes and a Fantasy* which contains a movement with only one note, the third movement of Ruth Crawford-Seeger's *String Quartet 1931* (later reorchestrated as *Andante for string orchestra*) in which the melody is created from an unchanging set of pitches through "dissonant dynamics" alone, and György Ligeti's *Aventures* in which recurring phonetics create the linear form.

See also

- Unified field
- Parsons code, a simple notation used to identify a piece of music through melodic motion—the motion of the pitch up and down.
- Appropriation (music)
- Klangfarbenmelodie
- *Musique concrète*
- Melodic patterns
- Sequence (music)

Melody from Anton Webern's *Variations for Orchestra*, Op. 30 (pp. 23-24)

Further reading

- Apel, Willi. *Harvard Dictionary of Music*, 2nd ed., p.517-19. [1]
- Edwards, Arthur C. *The Art of Melody*, p.xix-xxx. Includes "a catalog of sample definitions." [1]
- Holst, Imogen (1962/2008). *Tune*, Faber and Faber, London. ISBN 0571241980.
- Smits van Waesberghe, J. (1955). *A Textbook of Melody: A course in functional melodic analysis*, American Institute of Musicology. Includes "an attempt to formulate a theory of melody." [1]
- Szabolcsi, Bence (1965). *A History Of Melody*, Barrie and Rockliff, London.

External links

- MelodyCatcher: An accurate Search Engine for Melodies [3]
- Melodyhound: A Search Engine for Melodies [4]
- Tunespotting: Another Search Engine for Melodies [5]
- DoDoSoSo: Yet another Search Engine for Melodies [6]
- Carry A Tune Week List of Tunes [7]
- Access to IMSLP 12 collections of 1000 melodies arranged for solo instrument, [8] with melody-specific historical notes

References

[1] DeLone *et al.* (Eds.) (1975). *Aspects of Twentieth-Century Music*, chap. 4, p.270-301. Englewood Cliffs, New Jersey: Prentice-Hall. ISBN 0-13-049346-5.
[2] Narveson, Paul (1984). *Theory of Melody*. ISBN 0819138347.
[3] http://www.melodycatcher.com/
[4] http://www.melodyhound.com/
[5] http://www.tunespotting.com/
[6] http://www.everity.com/~gheller
[7] http://www.americanmusicpreservation.com/carryatuneweek.htm
[8] http://imslp.org/wiki/User:Clark_Kimberling/Historical_Notes_1

Miriam Stockley

Miriam Stockley (born on 15 April 1962 in Johannesburg, South Africa) is a British singer. Her work is influenced by the African music from her home country. She is married to Rod Houison. They have two children, Carly Houison and Leigh Brandon Houison.

At the age of eleven, Stockley and her older sister Avryl formed a group, the Stockley Sisters. Later in her life, she moved to London to further pursue her musical career. There, she contributed vocals to several albums, and TV commercials.

During the late eighties and early nineties, Stockley worked as a session singer for the UK song writing and production trio Stock, Aitken and Waterman. Stockley featured on tracks by the likes of

Miriam Stockley in Frankfurt am Main, December 22nd, 2006

Kylie Minogue, Jason Donovan and Sonia. Alongside fellow session singer Mae McKenna, Stockley is credited with being partly responsible for the distinctive Stock, Aitken and Waterman sound of the eighties.

Stockley provided backing vocals for the United Kingdom's Eurovision Song Contest entry on several occasions, most notably with Emma in 1990 and for Katrina and the Waves when they won in Dublin in 1997.

In 1991, Stockley became a part of the dance group Praise whose single "Only You" reached number four on the UK charts. A year later, the band, with Stockley on vocals, released their second single "Dream On". This failed to have the same success as their previous single and the band decided to call it a day. Later, she was the vocalist for Adiemus with Karl Jenkins.

Stockley has been featured on several film soundtracks, including her cover of the Rose Royce song "Wishing on a Star", which appears on *The 10th Kingdom* soundtrack.

Her song, "Perfect Day", was the theme song for the BBC children's programme The World of Peter Rabbit and Friends which ran from 1992 to 1995. The song was written by Colin Towns, and has since become unofficially known as 'Theme for the Lake District' by enthusiasts of the show. She has also appeared as a singer on several BBC school's programmes, most notably Look and Read.

In 2004, Yamaha released Vocaloid software that allows to synthetically create background vocals; one of the three available voices is based on material recorded by Miriam Stockley.

In December 2006 she contributed as a solo vocalist and as a co-vocalist with Mike Oldfield at the German Night of the Proms tour, consisting of 18 concerts. She also released her third solo album, a collection of rearranged classical standards entitled *Eternal* in 2006.

Discography

Full discography available at official website.

Solo albums

* *Miriam* (1999)
* *Second Nature* (2003)
* *Eternal* (2006)

Others

Backing vocals:

* Freddie Mercury & Montserrat Caballé - *Barcelona* (1988; backing vocals on "The Golden Boy")
* Brian May - *Back to the Light* (1992)
* Queen - Made in Heaven
* Mike Oldfield - *The Art in Heaven Concert* (2000; backing vocals and lead vocal on "Moonlight Shadow")
* Queen+: *The Freddie Mercury Tribute Concert* (DVD, 2002)
* Atlantis vs Avatar - *Fiji* (provided lead vocals; Inferno Records, 2000)
* The Lord of the Rings film trilogy

External links

* Official website [1]
* Official German Nokia - Night of the proms Website [2]
* Inferno Records [3]

References

[1] http://miriam.co.uk
[2] http://www.notp.com/?country=de&menuitem=1
[3] http://www.infernorecords.co.uk

NAMM Show

The **NAMM Show** is one of the largest music product trade shows in the world, its only major competition being the *Musik Messe* in Frankfurt. It is held every January in Anaheim, California, USA at the Anaheim Convention Center. The January 2008 show had 1,560 exhibitors and a record-breaking 88,100 attendees. The NAMM Show is not open to the general public, only to members of the music trade and/or those who have been invited.

The acronym **NAMM** originally stood for the **National Association of Music Merchants**, but has evolved from a national entity representing the interests of music products retailers to an international association including both commercial, retail members and affiliates. Therefore, the long form of the name is no longer used, and it is simply known as NAMM, the International Music Products Association.

Each Winter NAMM Show is heavily covered by the music-industry press. Thousands of new-product introductions and demonstrations are made at Winter NAMM, making attendance by trade journalists a necessity. Some publications, such as Music Trades[1], publish special NAMM issues for distribution at the show. NAMM also produces a show directory and a daily news magazine called *Upbeat*, for distribution at the convention center and at local hotels.

Among music-trade attendees, the NAMM Show is an exhausting ritual. The Anaheim Convention Center is one of the world's largest, and NAMM (since 2007) completely fills all the available exhibit space, necessitating miles of walking to cover all the exhibits. NAMM is also colloquially called "the world's loudest trade show"[2], although this monicker is also assigned to E3 and other shows involving entertainment. Despite severe regulations on the permissible noise level, sound level meters carried by NAMM personnel routinely exceed the 85 dBA maximum throughout most of the main exhibit hall, simply from the constant background noise. [3] [4]

The association's other show, Summer NAMM, was hosted by Austin, Texas at the end of July, 2007. The smaller summer show is approximately one third the size of the Winter NAMM Show and focuses more on industry meetings and professional development courses than products. Recent reports indicated that the Summer NAMM show has suffered from declining attendance, as it has been cycled between Nashville TN, Austin TX and Indianapolis IN during successive years. [5] [6]

2010 NAMM Show Performances

The 2010 NAMM Show featured performances from the following bands, finalists in the SchoolJam USA contest.

- Adrenaline
- Aftermath
- Almost Chaos
- Chasing the Skyline
- Crimson Fire
- Dance Over Anaheim
- High Tide
- Jaci And Those Guys
- Power Pirate
- Switchfits

External links

- Official NAMM website [7]
- Steve Vai on NAMM 2009 disussing the creation of Amplifiers for rock guitarists, Premier Guitar.com [8]
- [9] = covers 2010 NAMM in Anaheim, California
- About.com - NAMM Show 2007 Coverage [10]
- NAMM Oddities - archives going back to 1998 [11]
- SongPlacements.com partners with NAMM [12]
- Full Compass NAMM Report [13]
- Music Ramblings of NAMM [14]
- Harmony Central - NAMM Show coverage [15]
- Sonic State - NAMM Show coverage [16]
- Vintage Rock NAMM Show Reports [17]
- NAMM Show coverage at Synthtopia [18]
- NAMM Show Coverage (german) [19]

References

[1] http://www.musictrades.com/
[2] http://launch.groups.yahoo.com/group/bbshop/message/78189?l=1
[3] http://www.modernguitars.com/wolf/archives/002851.html
[4] http://www.behringer.com/01_news/events_namm08.cfm?lang=eng
[5] http://www.tracklists.ca/latest/2007-summer-namm-wraps-declining-attendance.html
[6] http://www.mi-pro.co.uk/news/28554/NAMM-back-at-Nashville
[7] http://www.namm.org/
[8] http://www.youtube.com/watch?v=f08_1nuOU_I
[9] http://www.guitarinternational.com/wpmu
[10] http://homerecording.about.com/od/namm2007coverage/NAMM_Show_2007_Coverage_Live_from_Anaheim.htm/
[11] http://www.otheroom.com/namm/
[12] http://songplacementscom.blogspot.com/2009/06/songplacements-to-host-showcase-at-namm.html
[13] http://www.fullcompass.com/news/170.html
[14] http://www.musicramble.com/news-from-namm-day-one/
[15] http://www.harmony-central.com/
[16] http://www.sonicstate.com/
[17] http://www.vintagerock.com/mainnamm.aspx
[18] http://www.synthtopia.com/content/tag/2010-namm-show/
[19] http://www.delamar.de/

Computer music

Computer music is a term that was originally used within academia to describe a field of study relating to the applications of computing technology in music composition; particularly that stemming from the Western art music tradition. It includes the theory and application of new and existing technologies in music, such as sound synthesis, digital signal processing, sound design, sonic diffusion, acoustics, and psychoacoustics. The field of computer music can trace its roots back to the origin of electronic music, and the very first experiments and innovations with electronic instruments at the turn of the 20th century. More recently, with the advent of personal computing, and the growth of home recording, the term computer music is now sometimes used to describe any music that has been created using computing technology.

History

Much of the work on computer music has drawn on the relationship between music theory and mathematics. The world's first computer to play music was CSIRAC which was designed and built by Trevor Pearcey and Maston Beard. Mathematician Geoff Hill programmed the CSIRAC to play popular musical melodies from the very early 1950s. In 1951 it publicly played the Colonel Bogey March[1] of which no known recordings exist. However, CSIRAC played standard repertoire and was not used to extend musical thinking or composition practice which is current computer-music practice.

The oldest known recordings of computer generated music were played by the Ferranti Mark 1 computer, a commercial version of the Baby Machine from the University of Manchester in the autumn of 1951. The music program was written by Christopher Strachey. During a session recorded by the BBC, the machine managed to work its way through "Baa Baa Black Sheep", "God Save the King" and part of "In the Mood".[2]

Two further major 1950s developments were the origins of digital sound synthesis by computer, and of algorithmic composition programs beyond rote playback. Max Mathews at Bell Laboratories developed the influential MUSIC I program and its descendents, further popularising computer music through a 1962 article in Science. Amongst other pioneers, the musical chemists Lejaren Hiller and Leonard Isaacson worked on a series of algorithmic composition experiments from 1956-9, manifested in the 1957 premiere of the *Illiac Suite* for string quartet.[3]

Early computer-music programs typically did not run in real time. Programs would run for hours or days, on multi-million-dollar computers, to generate a few minutes of music. John Chowning's work on FM synthesis from the 1960s to the 1970s, and the advent of inexpensive digital chips and microcomputers opened the door to real-time generation of computer music. By the early 1990s, the performance of microprocessor-based computers reached the point that real-time generation of computer music using more general programs and algorithms became possible.

Advances

Advances in computing power have dramatically affected the way computer music is generated and performed. Current-generation micro-computers are powerful enough to perform very sophisticated audio synthesis using a wide variety of algorithms and approaches. Computer music systems and approaches are now ubiquitous, and so firmly embedded in the process of creating music that we hardly give them a second thought: computer-based synthesizers, digital mixers, and effects units have become so commonplace that use of digital rather than analog technology to create and record music is the norm, rather than the exception.

Research

Despite the ubiquity of computer music in contemporary culture, there is considerable activity in the field of computer music, as researchers continue to pursue new and interesting computer-based synthesis, composition, and performance approaches.Throughout the world there are many organizations and institutions dedicated to the area of computer and electronic music study and research, including the ICMA (International Computer Music Association), IRCAM, GRAME, SEAMUS (Society for Electro Acoustic Music in the United States), and a great number of institutions of higher learning around the world.

Computer-generated music

Computer-generated music is music composed by, or with the extensive aid of, a computer. Although any music which uses computers in its composition or realisation is computer-generated to some extent, the use of computers is now so widespread (in the editing of pop songs, for instance) that the phrase computer-generated music is generally used to mean a kind of music which could not have been created *without* the use of computers.

We can distinguish two groups of computer-generated music: music in which a computer generated the score, which could be performed by humans, and music which is both composed and performed by computers.There is a large genre of music that is organized, synthesized, and created on computers.

Computer-generated scores for performance by human players

Many systems for generating musical scores actually existed well before the time of computers. One of these was Musikalisches Würfelspiel, a system which used throws of the dice to randomly select measures from a large collection of small phrases. When patched together, these phrases combined to create musical pieces which could be performed by human players. Although these works were not actually composed with a computer in the modern sense, it uses a rudimentary form of the random combinatorial techniques sometimes used in computer-generated composition.

The world's first digital computer music was generated in Australia by programmer Geoff Hill on the CSIRAC computer which was designed and built by Trevor Pearcey and Maston Beard, although it was only used to play standard tunes of the day. Subsequently, one of the first composers to write music with a computer was Iannis Xenakis. He wrote programs in the FORTRAN language that generated numeric data that he transcribed into scores to be played by traditional musical instruments. An example is *ST/48* of 1962. Although Xenakis could well have composed this music by hand, the intensity of the calculations needed to transform probabilistic mathematics into musical notation was best left to the number-crunching power of the computer.

Computers have also been used in an attempt to imitate the music of great composers of the past, such as Mozart. A present exponent of this technique is David Cope. He wrote computer programs that analyse works of other composers to produce new works in a similar style. He has used this program to great effect with composers such as Bach and Mozart (his program *Experiments in Musical Intelligence* is famous for creating "Mozart's 42nd Symphony"), and also within his own pieces, combining his own creations with that of the computer.

Music composed and performed by computers

Later, composers such as Gottfried Michael Koenig had computers generate the sounds of the composition as well as the score. Koenig produced algorithmic composition programs which were a generalisation of his own serial composition practice. This is not exactly similar to Xenakis' work as he used mathematical abstractions and examined how far he could explore these musically. Koenig's software translated the calculation of mathematical equations into codes which represented musical notation. This could be converted into musical notation by hand and then performed by human players. His programs Project 1 and Project 2 are examples of this kind of software. Later, he extended the same kind of principles into the realm of synthesis, enabling the computer to produce the sound directly. SSP is an example of a program which performs this kind of function. All of these programs were produced by Koenig at the Institute of Sonology in Utrecht, Holland in the 1970s.

Procedures such as those used by Koenig and Xenakis are still in use today. Since the invention of the MIDI system in the early 1980s, for example, some people have worked on programs which map MIDI notes to an algorithm and then can either output sounds or music through the computer's sound card or write an audio file for other programs to play.

Some of these simple programs are based on fractal geometry, and can map midi notes to specific fractals, or fractal equations. Although such programs are widely available and are sometimes seen as clever toys for the non-musician, some professional musicians have given them attention also. The resulting 'music' can be more like noise, or can sound quite familiar and pleasant. As with much algorithmic music, and algorithmic art in general, more depends on the way in which the parameters are mapped to aspects of these equations than on the equations themselves. Thus, for example, the same equation can be made to produce both a lyrical and melodic piece of music in the style of the mid-nineteenth century, and a fantastically dissonant cacophony more reminiscent of the avant-garde music of the 1950s and 1960s.

Other programs can map mathematical formulae and constants to produce sequences of notes. In this manner, an irrational number can give an infinite sequence of notes where each note is a digit in the decimal expression of that number. This sequence can in turn be a composition in itself, or simply the basis for further elaboration.

Operations such as these, and even more elaborate operations can also be performed in computer music programming languages such as Max/MSP, SuperCollider, Csound, Pure Data (Pd), Keykit, and ChucK. These programs now easily run on most personal computers, and are often capable of more complex functions than those which would have necessitated the most powerful mainframe computers several decades ago.

There exist programs that generate "human-sounding" melodies by using a vast database of phrases. One example is Band-in-a-Box, which is capable of creating jazz, blues and rock instrumental solos with almost no user interaction. Another is Impro-Visor, which uses a stochastic context-free grammar to generate phrases and complete solos.

Another 'cybernetic' approach to computer composition uses specialized hardware to detect external stimuli which are then mapped by the computer to realize the performance. Examples of this style of computer music can be found in the middle-80's work of David Rokeby (Very Nervous System) where audience/performer motions are 'translated' to MIDI segments. Computer controlled music is also found in the performance pieces by the Canadian composer Udo Kasemets (1919-) such as the Marce(ntennia)l Circus C(ag)elebrating Duchamp

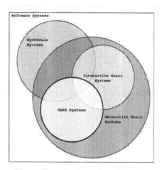

Diagram illustrating the position of CAAC in relation to other Generative music Systems

(1987), a realization of the Marcel Duchamp process piece Music Errata using an electric model train to collect a hopper-car of stones to be deposited on a drum wired to an Analog:Digital converter, mapping the stone impacts to a score display (performed in Toronto by pianist Gordon Monahan during the 1987 Duchamp Centennial), or his installations and performance works (e.g. Spectrascapes) based on his Geo(sono)scope (1986) 15x4-channel computer-controlled audio mixer. In these latter works, the computer generates sound-scapes from tape-loop sound samples, live shortwave or sine-wave generators.

Computer-Aided Algorithmic Composition

Computer-Aided Algorithmic Composition (CAAC, pronounced "sea-ack") is the implementation and use of algorithmic composition techniques in software. This label is derived from the combination of two labels, each too vague for continued use. The label "computer-aided composition" lacks the specificity of using generative algorithms. Music produced with notation or sequencing software could easily be considered computer-aided composition. The label "algorithmic composition" is likewise too broad, particularly in that it does not specify the use of a computer. The term computer-aided, rather than computer-assisted, is used in the same manner as Computer-Aided Design

Video-Driven Soundtrack Composer

A new concept insofar as the music generated is driven by an associated video. This process has been developed by Abaltat and is available as a commercial product Abaltat Muse for users to create their own music based on their video pictures or slideshows. The process involves an analysis of the color saturation in the pictures, together with the calculation of the duration of the footage. The user then chooses a style of music and a soundtrack is created using the video analysis information combined with the rules of the chosen musical style. The music generated is royalty-free since the user is the originator of the music.

Machine Improvisation

Machine Improvisation uses computer algorithms to create improvisation on existing music materials. This is usually done by sophisticated recombination of musical phrases extracted from existing music, either live or pre-recorded. In order to achieve credible improvisation in particular style, machine improvisation uses machine learning and pattern matching algorithms to analyze existing musical examples. The resulting patterns are then used to create new variations "in the style" of the original music, developing a notion of stylistic reinjection. This is different from other improvisation methods with computers that use algorithmic composition to generate new music without performing analysis of existing music examples.

Statistical style modeling

Style modeling implies building a computational representation of the musical surface that captures important stylistic features from data. Statistical approaches are used to capture the redundancies in terms of pattern dictionaries or repetitions, which are later recombined to generate new musical data. Style mixing can be realized by analysis of a database containing multiple musical examples in different styles. Machine Improvisation builds upon a long musical tradition of statistical modeling that began with Hiller and Isaacson's Illiac Suite in the 1950s and Xenakis' uses of Markov chains and stochastic processes. Modern methods include the use of lossless data compression for incremental parsing, Prediction Suffix Tree and string searching by factor oracle algorithm

Uses of Machine Improvisation

Machine Improvisation encourages musical creativity by providing automatic modeling and transformation structures for existing music. This creates a natural interface with the musician without need for coding musical algorithms. In live performance, the system re-injects the musician's material in several different ways, allowing a semantics-level representation of the session and a smart recombination and transformation of this material in real-time. In offline version, Machine Improvisation can be used to achieve style mixing, an approach inspired by Vannevar Bush's memex imaginary machine.

Implementations

Matlab implementation of the Factor Oracle machine improvisation can be found as part of Computer Audition toolbox.

OMax is a software environment developed in IRCAM. OMax uses OpenMusic and Max. It is based on researches on stylistic modeling carried out by Gerard Assayag and Shlomo Dubnov and on researches on improvisation with the computer by G. Assayag, M. Chemillier and G. Bloch (Aka the OMax Brothers) in the Ircam Music Representations group.

Musicians working with machine improvisation

Gerard Assayag (IRCAM, France), Tim Blackwell (Goldsmiths College, Great Brittan), George Bloch (Composer, France), Marc Chemiller (IRCAM/CNRS, France), Nick Collins (University of Sussex, UK) Shlomo Dubnov (Composer, Israel / USA), Mari Kimura (Juilliard, New York City), George Lewis (Columbia University, New York City), Bernard Lubat (Pianist, France), Joel Ryan (Institute of Sonology, Netherlands), Michel Waisvisz (STEIM, Netherlands), David Wessel (CNMAT, California), Michael Young (Goldsmiths College, Great Brittan), Pietro Grossi (CNUCE, Institute of the National Research Council, Pisa, Italy)

Live coding

Live coding[4] (sometimes known as 'interactive programming', 'on-the-fly programming',[5] 'just in time programming') is the name given to the process of writing software in realtime as part of a performance. Historically, similar techniques were used to produce early computer art, but recently it has been explored as a more rigorous alternative to laptop DJs who, live coders often feel, lack the charisma and pizzazz of musicians performing live.[6]

Generally, this practice stages a more general approach: one of interactive programming, of writing (parts of) programs while they run. Traditionally most computer music programs have tended toward the old write/compile/run model which evolved when computers were much less powerful. This approach has locked out code-level innovation by people whose programming skills are more modest. Some programs have gradually integrated real-time controllers and gesturing (for example, MIDI-driven software synthesis and parameter control). Until recently, however, the musician/composer rarely had the capability of real-time modification of program code itself. This legacy distinction is somewhat erased by languages such as ChucK, SuperCollider, and Impromptu.

TOPLAP, an ad-hoc conglomerate of artists interested in live coding was formed in 2004, and promotes the use, proliferation and exploration of a range of software, languages and techniques to implement live coding. This is a parallel and collaborative effort e.g. with research at the Princeton Sound Lab, the University of Cologne, and Computational Arts Research Group at Queensland University of Technology.

See also

- Acousmatic art
- Chiptune
- Comparison of audio synthesis environments
- Csound
- Digital audio workstation
- Digital synthesizer
- Electronic music
- Fast Fourier Transform
- Human-computer interaction
- Interactive music
- Laptronica
- Music information retrieval
- Music Macro Language
- Music notation software
- Music sequencer
- New interfaces for musical expression
- Physical modeling
- Sampling (music)
- sound synthesis
- Tracker

Further reading

- Ariza, C. 2005. "Navigating the Landscape of Computer-Aided Algorithmic Composition Systems: A Definition, Seven Descriptors, and a Lexicon of Systems and Research." In *Proceedings of the International Computer Music Conference*. San Francisco: International Computer Music Association. 765-772. Internet: http://www. flexatone.net/docs/nlcaacs.pdf
- Ariza, C. 2005. *An Open Design for Computer-Aided Algorithmic Music Composition: athenaCL*. Ph.D. Dissertation, New York University. Internet: http://www.flexatone.net/docs/odcaamca.pdf
- Berg, P. 1996. "Abstracting the future: The Search for Musical Constructs" *Computer Music Journal* 20(3): 24-27.
- Boulanger, Richard, ed (March 6, 2000). *The Csound Book: Perspectives in Software Synthesis, Sound Design, Signal Processing, and Programming* [7]. The MIT Press. pp. 740. ISBN 0262522616. Retrieved 3 Oct 2009.
- Chadabe, Joel. 1997. *Electric Sound: The Past and Promise of Electronic Music*. Upper Saddle River, New Jersey: Prentice Hall.
- Chowning, John. 1973. "The Synthesis of Complex Audio Spectra by Means of Frequency Modulation". *Journal of the Audio Engineering Society* 21, no. 7:526–34.
- Collins, Nick (2009). *Introduction to Computer Music*. Chichester: Wiley. ISBN 9780470714553.
- Dodge, Charles; Jerse (1997). *Computer Music: Synthesis, Composition and Performance*. Thomas A. (2nd ed.). New York: Schirmer Books. pp. 453. ISBN 0-02-864682-7.
- Heifetz, Robin (1989). *On the Wires of Our Nerves*. Lewisburg Pa.: Bucknell University Press. ISBN 0838751555.
- Manning, Peter (2004). *Electronic and Computer Music* (revised and expanded ed.). Oxford Oxfordshire: Oxford University Press. ISBN 0195170857.
- Roads, Curtis (1994). *The Computer Music Tutorial*. Cambridge: MIT Press. ISBN 0262680823.
- Supper, M. 2001. "A Few Remarks on Algorithmic Composition." *Computer Music Journal* 25(1): 48-53.
- Xenakis, Iannis (2001). *Formalized Music: Thought and Mathematics in Composition*. Harmonologia Series No. 6. Hillsdale, NY: Pendragon Pr. ISBN 1576470792.

External links

- Computer Music Forum [8]

Software environments

- AC Toolbox [9]
- Bol Processor [10]
- ChucK [11], a strongly-timed, concurrent, and on-the-fly language
- Common Music [12], a music composition environment that produces sound by transforming a high-level representation of musical structure into a variety of control protocols for sound synthesis and display.
- Csound [13]
- fluxus [14] livecoding and playing/learning environment for 3D graphics and games based on Scheme
- impromptu audiovisual livecoding environment
- Impro-Visor improvisation instruction, with automatic melody generation
- Max/MSP, a graphical development environment for music and multimedia, invented at IRCAM, named after Max Mathews, and developed and maintained by San Francisco-based software company Cycling '74
- MEAPsoft [15] descriptor based audio segmentation and re-arrangement
- KeyKit [16]
- OMax software [17]
- OpenMusic [18]
- Pd [19]
- Processing [20]
- PWGL [21], a free cross-platform visual language based on Common Lisp, CLOS and OpenGL, specialized in computer aided composition and sound synthesis
- SuperCollider [22]
- Symbolic Composer [23]

Articles

- Computer Generated Music Composition [24] thesis by Chong Yu (MIT 1996)
- Computer-aided Composition [25] article by Karlheinz Essl (1991)
- G. Assayag, S. Dubnov « Using Factor Oracles for machine Improvisation », Soft Computing, vol. 8, n° 9, Septembre, 2004 [26]
- S. Dubnov et al. Using machine-learning methods for musical style modeling, IEEE Computer, Oct. 2003 [27]
- G. Lewis, Too Many Notes: Computers, Complexity and Culture in Voyager, Leonardo Music Journal 10 (2000) 33-39 [28]
- S. Dubnov, Stylistic randomness: about composing NTrope Suite, Organised Sound, Volume 4 , Issue 2 (June 1999) [29]

Archives

- algorithmic.net [30] - a lexicon of systems and research in computer aided algorithmic composition
- doornbusch.net/chronology [31] - a chronology of computer music and related events

Works composed by computers for human performance

- Illiac Suite [32] for string quartet, by Lejaren A. Hiller and Leonard Isaacson (1957)
- Übung, 3 Asko Pieces, Beitrag [33] (amongst others) by G.M. Koenig

Computer-generated compositions performed by computers

- Lexikon-Sonate: [34] Karlheinz Essl's algorithmic composition environment
- Metamath Music [35] Music generated from mathematical proofs
- CodeSounding [36] Sonification of java source code structures, obtained by post-processing the source files. Runtime sounds are a function of how the source code of the running program was structured
- Virtual Music Composer [37] This software works as a composer, not as a tool for composing
- Fractal Tune Smithy [38] Computer generated music based on a similar idea to the Koch snowflake, with many examples of tunes you can make
- ALICE [39] A software that can improvise in real-time with a human player using an Artificial neural network
- [[viral symphOny [40]]] created using computer virus software by Joseph Nechvatal

References

[1] Doornbusch, Paul. "The Music of CSIRAC" (http://www.csse.unimelb.edu.au/dept/about/csirac/music/introduction.html). *Melbourne School of Engineering, Department of Computer Science and Software Engineering.* .

[2] Fildes, Jonathan (June 17, 2008). "'Oldest' computer music unveiled" (http://news.bbc.co.uk/2/hi/technology/7458479.stm). *News.bbc.co.uk.* . Retrieved 2008-06-17.

[3] Lejaren Hiller and Leonard Isaacson. Experimental Music: Composition with an Electronic Computer. Greenwood Press, 1959.

[4] Collins, N., McLean, A., Rohrhuber, J. & Ward, A. (2003), "Live Coding Techniques for Laptop Performance", *Organised Sound* 8(3):321–30. doi: 10.1017/S135577180300030X (http://dx.doi.org/10.1017/S135577180300030X)

[5] Wang G. & Cook P. (2004) "On-the-fly Programming: Using Code as an Expressive Musical Instrument" (http://soundlab.cs.princeton.edu/publications/on-the-fly_nime2004.pdf), In *Proceedings of the 2004 International Conference on New Interfaces for Musical Expression (NIME)* (New York: NIME, 2004).

[6] Collins, N. (2003) "Generative Music and Laptop Performance", *Contemporary Music Review* 22(4):67–79.

[7] http://csounds.com/shop/csound-book

[8] http://computermusicforum.mforos.com

[9] http://www.koncon.nl/downloads/ACToolbox/

[10] http://bolprocessor.sourceforge.net/

[11] http://chuck.cs.princeton.edu/

[12] http://commonmusic.sourceforge.net/

[13] http://www.csounds.com/

[14] http://www.pawfal.org/fluxus/

[15] http://www.meapsoft.org/

[16] http://www.nosuch.com/keykit

[17] http://recherche.ircam.fr/equipes/repmus/OMax/

[18] http://recherche.ircam.fr/equipes/repmus/OpenMusic/

[19] http://www.puredata.info/

[20] http://processing.org/

[21] http://www2.siba.fi/PWGL/

[22] http://supercollider.sourceforge.net/

[23] http://www.symboliccomposer.com/

[24] http://home.comcast.net/~chtongyu/Thesis.html

[25] http://www.essl.at/bibliogr/cac.html

[26] http://mediatheque.ircam.fr/articles/textes/Assayag04a/

[27] http://ieeexplore.ieee.org/xpls/abs_all.jsp?arnumber=1236474

[28] http://muse.jhu.edu/demo/leonardo_music_journal/v010/10.1lewis.html

[29] http://journals.cambridge.org/action/displayIssue?jid=OSO&volumeId=4&issueId=02#

[30] http://www.algorithmic.net

[31] http://www.doornbusch.net/chronology

[32] http://emfinstitute.emf.org/exhibits/illiacsuite.html

[33] http://www.koenigproject.nl

[34] http://www.essl.at/works/Lexikon-Sonate.html

[35] http://us.metamath.org/mpegif/mmmusic.html

[36] http://www.codesounding.org/indexeng.html

[37] http://www.lvbsx.com

[38] http://www.robertinventor.com/software/tunesmithy/tune_smithying.htm

[39] http://www.brandmaier.de/alice/

[40] http://www.ubu.com/sound/nechvatal.html

Speech synthesis

Speech synthesis is the artificial production of human speech. A computer system used for this purpose is called a **speech synthesizer**, and can be implemented in software or hardware. A **text-to-speech** (**TTS**) system converts normal language text into speech; other systems render symbolic linguistic representations like phonetic transcriptions into speech.[1]

Synthesized speech can be created by concatenating pieces of recorded speech that are stored in a database. Systems differ in the size of the stored speech units; a system that stores phones or diphones provides the largest output range, but may lack clarity. For specific usage domains, the storage of entire words or sentences allows for high-quality output. Alternatively, a synthesizer can incorporate a model of the vocal tract and other human voice characteristics to create a completely "synthetic" voice output.[2]

The quality of a speech synthesizer is judged by its similarity to the human voice and by its ability to be understood. An intelligible text-to-speech program allows people with visual impairments or reading disabilities to listen to written works on a home computer. Many computer operating systems have included speech synthesizers since the early 1980s.

Overview of text processing

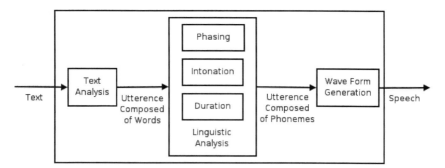

A text-to-speech system (or "engine") is composed of two parts: a front-end and a back-end. The front-end has two major tasks. First, it converts raw text containing symbols like numbers and abbreviations into the equivalent of written-out words. This process is often called *text normalization*, *pre-processing*, or *tokenization*. The front-end then assigns phonetic transcriptions to each word, and divides and marks the text into prosodic units, like phrases, clauses, and sentences. The process of assigning phonetic transcriptions to words is called *text-to-phoneme* or *grapheme-to-phoneme* conversion.[3] Phonetic transcriptions and prosody information together make up the symbolic linguistic representation that is output by the front-end. The back-end—often referred to as the *synthesizer*—then converts the symbolic linguistic representation into sound.

History

Long before electronic signal processing was invented, there were those who tried to build machines to create human speech. Some early legends of the existence of "speaking heads" involved Gerbert of Aurillac (d. 1003 AD), Albertus Magnus (1198–1280), and Roger Bacon (1214–1294).

In 1779, the Danish scientist Christian Kratzenstein, working at the Russian Academy of Sciences, built models of the human vocal tract that could produce the five long vowel sounds (in International Phonetic Alphabet notation, they are [aː], [eː], [iː], [oː] and [uː]).[4] This was followed by the bellows-operated "acoustic-mechanical speech machine" by Wolfgang von Kempelen of Vienna, Austria, described in a 1791 paper.[5] This machine added models of the tongue and lips, enabling it to produce consonants as well as vowels. In 1837, Charles Wheatstone produced a "speaking machine" based on von Kempelen's design, and in 1857, M. Faber built the "Euphonia". Wheatstone's design was resurrected in 1923 by Paget.[6]

In the 1930s, Bell Labs developed the VOCODER, a keyboard-operated electronic speech analyzer and synthesizer that was said to be clearly intelligible. Homer Dudley refined this device into the VODER, which he exhibited at the 1939 New York World's Fair.

The Pattern playback was built by Dr. Franklin S. Cooper and his colleagues at Haskins Laboratories in the late 1940s and completed in 1950. There were several different versions of this hardware device but only one currently survives. The machine converts pictures of the acoustic patterns of speech in the form of a spectrogram back into sound. Using this device, Alvin Liberman and colleagues were able to discover acoustic cues for the perception of phonetic segments (consonants and vowels).

Early electronic speech synthesizers sounded robotic and were often barely intelligible. The quality of synthesized speech has steadily improved, but output from contemporary speech synthesis systems is still clearly distinguishable from actual human speech.

Electronic devices

The first computer-based speech synthesis systems were created in the late 1950s, and the first complete text-to-speech system was completed in 1968. In 1961, physicist John Larry Kelly, Jr and colleague Louis Gerstman[7] used an IBM 704 computer to synthesize speech, an event among the most prominent in the history of Bell Labs. Kelly's voice recorder synthesizer (vocoder) recreated the song "Daisy Bell", with musical accompaniment from Max Mathews. Coincidentally, Arthur C. Clarke was visiting his friend and colleague John Pierce at the Bell Labs Murray Hill facility. Clarke was so impressed by the demonstration that he used it in the climactic scene of his screenplay for his novel *2001: A Space Odyssey*,[8] where the HAL 9000 computer sings the same song as it is being put to sleep by astronaut Dave Bowman.[9] Despite the success of purely electronic speech synthesis, research is still being conducted into mechanical speech synthesizers.[10]

Synthesizer technologies

The most important qualities of a speech synthesis system are *naturalness* and *Intelligibility*. Naturalness describes how closely the output sounds like human speech, while intelligibility is the ease with which the output is understood. The ideal speech synthesizer is both natural and intelligible. Speech synthesis systems usually try to maximize both characteristics.

The two primary technologies for generating synthetic speech waveforms are *concatenative synthesis* and *formant synthesis*. Each technology has strengths and weaknesses, and the intended uses of a synthesis system will typically determine which approach is used.

Concatenative synthesis

Concatenative synthesis is based on the concatenation (or stringing together) of segments of recorded speech. Generally, concatenative synthesis produces the most natural-sounding synthesized speech. However, differences between natural variations in speech and the nature of the automated techniques for segmenting the waveforms sometimes result in audible glitches in the output. There are three main sub-types of concatenative synthesis.

Unit selection synthesis

Unit selection synthesis uses large databases of recorded speech. During database creation, each recorded utterance is segmented into some or all of the following: individual phones, diphones, half-phones, syllables, morphemes, words, phrases, and sentences. Typically, the division into segments is done using a specially modified speech recognizer set to a "forced alignment" mode with some manual correction afterward, using visual representations such as the waveform and spectrogram.[11] An index of the units in the speech database is then created based on the segmentation and acoustic parameters like the fundamental frequency (pitch), duration, position in the syllable, and neighboring phones. At runtime, the desired target utterance is created by determining the best chain of candidate units from the database (unit selection). This process is typically achieved using a specially weighted decision tree.

Unit selection provides the greatest naturalness, because it applies only a small amount of digital signal processing (DSP) to the recorded speech. DSP often makes recorded speech sound less natural, although some systems use a small amount of signal processing at the point of concatenation to smooth the waveform. The output from the best unit-selection systems is often indistinguishable from real human voices, especially in contexts for which the TTS system has been tuned. However, maximum naturalness typically require unit-selection speech databases to be very large, in some systems ranging into the gigabytes of recorded data, representing dozens of hours of speech.[12] Also, unit selection algorithms have been known to select segments from a place that results in less than ideal synthesis (e.g. minor words become unclear) even when a better choice exists in the database.[13]

Diphone synthesis

Diphone synthesis uses a minimal speech database containing all the diphones (sound-to-sound transitions) occurring in a language. The number of diphones depends on the phonotactics of the language: for example, Spanish has about 800 diphones, and German about 2500. In diphone synthesis, only one example of each diphone is contained in the speech database. At runtime, the target prosody of a sentence is superimposed on these minimal units by means of digital signal processing techniques such as linear predictive coding, PSOLA[14] or MBROLA.[15] The quality of the resulting speech is generally worse than that of unit-selection systems, but more natural-sounding than the output of formant synthesizers. Diphone synthesis suffers from the sonic glitches of concatenative synthesis and the robotic-sounding nature of formant synthesis, and has few of the advantages of either approach other than small size. As such, its use in commercial applications is declining, although it continues to be used in research because there are a number of freely available software implementations.

Domain-specific synthesis

Domain-specific synthesis concatenates prerecorded words and phrases to create complete utterances. It is used in applications where the variety of texts the system will output is limited to a particular domain, like transit schedule announcements or weather reports.[16] The technology is very simple to implement, and has been in commercial use for a long time, in devices like talking clocks and calculators. The level of naturalness of these systems can be very high because the variety of sentence types is limited, and they closely match the prosody and intonation of the original recordings.

Because these systems are limited by the words and phrases in their databases, they are not general-purpose and can only synthesize the combinations of words and phrases with which they have been preprogrammed. The blending of words within naturally spoken language however can still cause problems unless the many variations are taken into account. For example, in non-rhotic dialects of English the "r" in words like "clear" /ˈkliːə/ is usually only pronounced when the following word has a vowel as its first letter (e.g. "clear out" is realized as /ˌkliːəɹˈaʊt/). Likewise in French, many final consonants become no longer silent if followed by a word that begins with a vowel, an effect called liaison. This alternation cannot be reproduced by a simple word-concatenation system, which would require additional complexity to be context-sensitive.

Formant synthesis

Formant synthesis does not use human speech samples at runtime. Instead, the synthesized speech output is created using an acoustic model. Parameters such as fundamental frequency, voicing, and noise levels are varied over time to create a waveform of artificial speech. This method is sometimes called *rules-based synthesis*; however, many concatenative systems also have rules-based components.

Many systems based on formant synthesis technology generate artificial, robotic-sounding speech that would never be mistaken for human speech. However, maximum naturalness is not always the goal of a speech synthesis system, and formant synthesis systems have advantages over concatenative systems. Formant-synthesized speech can be reliably intelligible, even at very high speeds, avoiding the acoustic glitches that commonly plague concatenative systems. High-speed synthesized speech is used by the visually impaired to quickly navigate computers using a screen reader. Formant synthesizers are usually smaller programs than concatenative systems because they do not have a database of speech samples. They can therefore be used in embedded systems, where memory and microprocessor power are especially limited. Because formant-based systems have complete control of all aspects of the output speech, a wide variety of prosodies and intonations can be output, conveying not just questions and statements, but a variety of emotions and tones of voice.

Examples of non-real-time but highly accurate intonation control in formant synthesis include the work done in the late 1970s for the Texas Instruments toy Speak & Spell, and in the early 1980s Sega arcade machines.[17] Creating proper intonation for these projects was painstaking, and the results have yet to be matched by real-time text-to-speech interfaces.[18]

Articulatory synthesis

Articulatory synthesis refers to computational techniques for synthesizing speech based on models of the human vocal tract and the articulation processes occurring there. The first articulatory synthesizer regularly used for laboratory experiments was developed at Haskins Laboratories in the mid-1970s by Philip Rubin, Tom Baer, and Paul Mermelstein. This synthesizer, known as ASY, was based on vocal tract models developed at Bell Laboratories in the 1960s and 1970s by Paul Mermelstein, Cecil Coker, and colleagues.

Until recently, articulatory synthesis models have not been incorporated into commercial speech synthesis systems. A notable exception is the NeXT-based system originally developed and marketed by Trillium Sound Research, a spin-off company of the University of Calgary, where much of the original research was conducted. Following the demise of the various incarnations of NeXT (started by Steve Jobs in the late 1980s and merged with Apple

Computer in 1997), the Trillium software was published under the GNU General Public License, with work continuing as gnuspeech. The system, first marketed in 1994, provides full articulatory-based text-to-speech conversion using a waveguide or transmission-line analog of the human oral and nasal tracts controlled by Carré's "distinctive region model".

HMM-based synthesis

HMM-based synthesis is a synthesis method based on hidden Markov models, also called Statistical Parametric Synthesis. In this system, the frequency spectrum (vocal tract), fundamental frequency (vocal source), and duration (prosody) of speech are modeled simultaneously by HMMs. Speech waveforms are generated from HMMs themselves based on the maximum likelihood criterion.[19]

Sinewave synthesis

Sinewave synthesis is a technique for synthesizing speech by replacing the formants (main bands of energy) with pure tone whistles.[20]

Challenges

Text normalization challenges

The process of normalizing text is rarely straightforward. Texts are full of heteronyms, numbers, and abbreviations that all require expansion into a phonetic representation. There are many spellings in English which are pronounced differently based on context. For example, "My latest project is to learn how to better project my voice" contains two pronunciations of "project".

Most text-to-speech (TTS) systems do not generate semantic representations of their input texts, as processes for doing so are not reliable, well understood, or computationally effective. As a result, various heuristic techniques are used to guess the proper way to disambiguate homographs, like examining neighboring words and using statistics about frequency of occurrence.

Recently TTS systems have begun to use HMMs (discussed above) to generate "parts of speech" to aid in disambiguating homographs. This technique is quite successful for many cases such as whether "read" should be pronounced as "red" implying past tense, or as "reed" implying present tense. Typical error rates when using HMMs in this fashion are usually below five percent. These techniques also work well for most European languages, although access to required training corpora is frequently difficult in these languages.

Deciding how to convert numbers is another problem that TTS systems have to address. It is a simple programming challenge to convert a number into words (at least in English), like "1325" becoming "one thousand three hundred twenty-five." However, numbers occur in many different contexts; "1325" may also be read as "one three two five", "thirteen twenty-five" or "thirteen hundred and twenty five". A TTS system can often infer how to expand a number based on surrounding words, numbers, and punctuation, and sometimes the system provides a way to specify the context if it is ambiguous[21] . Roman numerals can also be read differently depending on context. For example "Henry VIII" reads as "Henry the Eighth", while "Chapter VIII" reads as "Chapter Eight".

Similarly, abbreviations can be ambiguous. For example, the abbreviation "in" for "inches" must be differentiated from the word "in", and the address "12 St John St." uses the same abbreviation for both "Saint" and "Street". TTS systems with intelligent front ends can make educated guesses about ambiguous abbreviations, while others provide the same result in all cases, resulting in nonsensical (and sometimes comical) outputs.

Text-to-phoneme challenges

Speech synthesis systems use two basic approaches to determine the pronunciation of a word based on its spelling, a process which is often called text-to-phoneme or grapheme-to-phoneme conversion (phoneme is the term used by linguists to describe distinctive sounds in a language). The simplest approach to text-to-phoneme conversion is the dictionary-based approach, where a large dictionary containing all the words of a language and their correct pronunciations is stored by the program. Determining the correct pronunciation of each word is a matter of looking up each word in the dictionary and replacing the spelling with the pronunciation specified in the dictionary. The other approach is rule-based, in which pronunciation rules are applied to words to determine their pronunciations based on their spellings. This is similar to the "sounding out", or synthetic phonics, approach to learning reading.

Each approach has advantages and drawbacks. The dictionary-based approach is quick and accurate, but completely fails if it is given a word which is not in its dictionary. As dictionary size grows, so too does the memory space requirements of the synthesis system. On the other hand, the rule-based approach works on any input, but the complexity of the rules grows substantially as the system takes into account irregular spellings or pronunciations. (Consider that the word "of" is very common in English, yet is the only word in which the letter "f" is pronounced [v].) As a result, nearly all speech synthesis systems use a combination of these approaches.

Languages with a phonemic orthography have a very regular writing system, and the prediction of the pronunciation of words based on their spellings is quite successful. Speech synthesis systems for such languages often use the rule-based method extensively, resorting to dictionaries only for those few words, like foreign names and borrowings, whose pronunciations are not obvious from their spellings. On the other hand, speech synthesis systems for languages like English, which have extremely irregular spelling systems, are more likely to rely on dictionaries, and to use rule-based methods only for unusual words, or words that aren't in their dictionaries.

Evaluation challenges

The consistent evaluation of speech synthesis systems may be difficult because of a lack of universally agreed objective evaluation criteria. Different organizations often use different speech data. The quality of speech synthesis systems also depends to a large degree on the quality of the production technique (which may involve analogue or digital recording) and on the facilities used to replay the speech. Evaluating speech synthesis systems has therefore often been compromised by differences between production techniques and replay facilities.

Recently, however, some researchers have started to evaluate speech synthesis systems using a common speech dataset.[22] .

Prosodics and emotional content

A recent study reported in the journal "**Speech Communication**" by Amy Drahota and colleagues at the University of Portsmouth, UK, reported that listeners to voice recordings could determine, at better than chance levels, whether or not the speaker was smiling.[23] It was suggested that identification of the vocal features which signal emotional content may be used to help make synthesized speech sound more natural.

Dedicated hardware

- Votrax
  - SC-01A (analog formant)
  - SC-02 / SSI-263 / "Arctic 263"
- General Instruments SP0256-AL2 (CTS256A-AL2, MEA8000)
- Magnevation SpeakJet (www.speechchips.com TTS256)
- Savage Innovations SoundGin
- National Semiconductor DT1050 Digitalker (Mozer)

- Silicon Systems SSI 263 (analog formant)
- Texas Instruments
  - TMS5110A (LPC) (obsolete?)
  - TMS5200 (obsolete?)
- Oki Semiconductor
  - ML22825 (ADPCM)
  - ML22573 (HQADPCM)
- Toshiba T6721A
- Philips PCF8200

Computer operating systems or outlets with speech synthesis

Apple

The first speech system integrated into an operating system was Apple Computer's MacInTalk in 1984. Since the 1980s Macintosh Computers offered text to speech capabilities through The MacinTalk software. In the early 1990s Apple expanded its capabilities offering system wide text-to-speech support. With the introduction of faster PowerPC-based computers they included higher quality voice sampling. Apple also introduced speech recognition into its systems which provided a fluid command set. More recently, Apple has added sample-based voices. Starting as a curiosity, the speech system of Apple Macintosh has evolved into a cutting edge fully-supported program, PlainTalk, for people with vision problems. VoiceOver was included in Mac OS X Tiger and more recently Mac OS X Leopard. The voice shipping with Mac OS X 10.5 ("Leopard") is called "Alex" and features the taking of realistic-sounding breaths between sentences, as well as improved clarity at high read rates. The operating system also includes say, a command-line based application that converts text to audible speech.

AmigaOS

The second operating system with advanced speech synthesis capabilities was AmigaOS, introduced in 1985. The voice synthesis was licensed by Commodore International from a third-party software house (Don't Ask Software, now Softvoice, Inc.) and it featured a complete system of voice emulation, with both male and female voices and "stress" indicator markers, made possible by advanced features of the Amiga hardware audio chipset.[24] It was divided into a narrator device and a translator library. Amiga Speak Handler featured a text-to-speech translator. AmigaOS considered speech synthesis a virtual hardware device, so the user could even redirect console output to it. Some Amiga programs, such as word processors, made extensive use of the speech system.

Microsoft Windows

Modern Windows systems use SAPI4- and SAPI5-based speech systems that include a speech recognition engine (SRE). SAPI 4.0 was available on Microsoft-based operating systems as a third-party add-on for systems like Windows 95 and Windows 98. Windows 2000 added a speech synthesis program called Narrator, directly available to users. All Windows-compatible programs could make use of speech synthesis features, available through menus once installed on the system. Microsoft Speech Server is a complete package for voice synthesis and recognition, for commercial applications such as call centers.

Text-to-Speech (TTS) capabilities for a computer refers to the ability to play back text in a spoken voice. **TTS** is the ability of the operating system to play back printed text as spoken words[25].

An internal (installed with the operating system) driver (called a TTS engine): recognizes the text and using a synthesized voice (chosen from several pre-generated voices) speaks the written text. Additional engines (often use a certain jargon or vocabulary) are also available through third-party manufacturers[25].

Android

Version 1.6 of Android added support for speech synthesis (TTS) [26] .

Internet

Currently, there are a number of applications, plugins and gadgets that can read messages directly from an e-mail client and web pages from a web browser. Some specialized software can narrate RSS-feeds. On one hand, online RSS-narrators simplify information delivery by allowing users to listen to their favourite news sources and to convert them to podcasts. On the other hand, on-line RSS-readers are available on almost any PC connected to the Internet. Users can download generated audio files to portable devices, e.g. with a help of podcast receiver, and listen to them while walking, jogging or commuting to work.

A growing field in internet based TTS is web-based assistive technology, e.g. 'Browsealoud' from a UK company. It can deliver TTS functionality to anyone (for reasons of accessibility, convenience, entertainment or information) with access to a web browser. Additionally SPEAK.TO.ME [27] from Oxford Information Laboratories is capable of delivering text to speech through any browser without the need to download any special applications, and includes smart delivery technology to ensure only what is seen is spoken and the content is logically pathed.

Others

- Some models of Texas Instruments home computers produced in 1979 and 1981 (Texas Instruments TI-99/4 and TI-99/4A) were capable of text-to-phoneme synthesis or reciting complete words and phrases (text-to-dictionary), using a very popular Speech Synthesizer peripheral. TI used a proprietary codec to embed complete spoken phrases into applications, primarily video games.[28]
- IBM's OS/2 Warp 4 included VoiceType, a precursor to IBM ViaVoice.
- Systems that operate on free and open source software systems including GNU/Linux are various, and include open-source programs such as the Festival Speech Synthesis System which uses diphone-based synthesis (and can use a limited number of MBROLA voices), and gnuspeech which uses articulatory synthesis[29] from the Free Software Foundation. Other proprietary software also runs on GNU/Linux.
- Several commercial companies are also developing speech synthesis systems : ABC2MP3 [30], Acapela Group [31], AT&T, Cepstral, CereProc [32], DECtalk, IBM ViaVoice, IVONA TTS [33], Loquendo TTS [34], NeoSpeech TTS [35], Nuance Communications, Orpheus [36], SVOX, YAKiToMe! [37], Voice on the Go [38], and Voxette [39].
- Companies which developed speech synthesis systems but which are no longer in this business include BeST Speech (bought by L&H), Eloquent Technology (bought by SpeechWorks), Lernout & Hauspie (bought by Nuance), SpeechWorks (bought by Nuance), Rhetorical Systems (bought by Nuance).

Speech synthesis markup languages

A number of markup languages have been established for the rendition of text as speech in an XML-compliant format. The most recent is Speech Synthesis Markup Language (SSML), which became a W3C recommendation in 2004. Older speech synthesis markup languages include Java Speech Markup Language (JSML) and SABLE. Although each of these was proposed as a standard, none of them has been widely adopted.

Speech synthesis markup languages are distinguished from dialogue markup languages. VoiceXML, for example, includes tags related to speech recognition, dialogue management and touchtone dialing, in addition to text-to-speech markup.

Applications

Accessibility

Speech synthesis has long been a vital assistive technology tool and its application in this area is significant and widespread. It allows environmental barriers to be removed for people with a wide range of disabilities. The longest application has been in the use of screenreaders for people with visual impairment, but text-to-speech systems are now commonly used by people with dyslexia and other reading difficulties as well as by pre-literate youngsters. They are also frequently employed to aid those with severe speech impairment usually through a dedicated voice output communication aid.

News service

Sites such as Ananova and YAKiToMe! [37] have used speech synthesis to convert written news to audio content, which can be used for mobile applications.

Entertainment

Speech synthesis techniques are used as well in the entertainment productions such as games, anime and similar. In 2007, Animo Limited announced the development of a software application package based on its speech synthesis software FineSpeech, explicitly geared towards customers in the entertainment industries, able to generate narration and lines of dialogue according to user specifications.[40] The application reached maturity in 2008, when NEC Biglobe announced a web service that allows users to create phrases from the voices of Code Geass: Lelouch of the Rebellion R2 characters.[41]

TTS applications such as YAKiToMe! [37] and Speakonia are often used to add synthetic voices to YouTube videos for comedic effect, as in Barney Bunch videos. YAKiToMe! [37] is also used to convert entire books for personal podcasting purposes, RSS feeds and web pages for news stories, and educational texts for enhanced learning.

Software such as Vocaloid can generate singing voices via lyrics and melody. This is also the aim of the Singing Computer project (which uses the GPL software Lilypond and Festival) to help blind people check their lyric input.[42]

Document eBook speaker

Android application MultiReader [43] speaks several document formats using TTS.

See also

References

[44]

[1] Jonathan Allen, M. Sharon Hunnicutt, Dennis Klatt, *From Text to Speech: The MITalk system*. Cambridge University Press: 1987. ISBN 0521306418

[2] Rubin, P., Baer, T., & Mermelstein, P. (1981). An articulatory synthesizer for perceptual research. *Journal of the Acoustical Society of America*, 70, 321-328.

[3] P. H. Van Santen, Richard William Sproat, Joseph P. Olive, and Julia Hirschberg, *Progress in Speech Synthesis*. Springer: 1997. ISBN 0387947019

[4] History and Development of Speech Synthesis (http://www.acoustics.hut.fi/publications/files/theses/lemmetty_mst/chap2.html), Helsinki University of Technology, Retrieved on November 4, 2006

[5] *Mechanismus der menschlichen Sprache nebst der Beschreibung seiner sprechenden Maschine* ("Mechanism of the human speech with description of its speaking machine," J. B. Degen, Wien).

[6] Mattingly, Ignatius G. Speech synthesis for phonetic and phonological models. In Thomas A. Sebeok (Ed.), *Current Trends in Linguistics, Volume 12, Mouton*, The Hague, pp. 2451-2487, 1974.

[7] http://query.nytimes.com/search/query?ppds=per&v1=GERSTMAN%2C%20LOUIS&sort=newest NY Times obituary for Louis Gerstman.

[8] Arthur C. Clarke online Biography (http://www.lsi.usp.br/~rbianchi/clarke/ACC.Biography.html)

[9] Bell Labs: Where "HAL" First Spoke (Bell Labs Speech Synthesis website) (http://www.bell-labs.com/news/1997/march/5/2.html)

[10] Anthropomorphic Talking Robot Waseda-Talker Series (http://www.takanishi.mech.waseda.ac.jp/research/voice/)

[11] Alan W. Black, Perfect synthesis for all of the people all of the time. IEEE TTS Workshop 2002. (http://www.cs.cmu.edu/~awb/papers/IEEE2002/allthetime/allthetime.html)

[12] John Kominek and Alan W. Black. (2003). CMU ARCTIC databases for speech synthesis. CMU-LTI-03-177. Language Technologies Institute, School of Computer Science, Carnegie Mellon University.

[13] Julia Zhang. Language Generation and Speech Synthesis in Dialogues for Language Learning, masters thesis, http://groups.csail.mit.edu/sls/publications/2004/zhang_thesis.pdf Section 5.6 on page 54.

[14] PSOLA Synthesis (http://www.fon.hum.uva.nl/praat/manual/PSOLA.html)

[15] T. Dutoit, V. Pagel, N. Pierret, F. Bataiile, O. van der Vrecken. The MBROLA Project: Towards a set of high quality speech synthesizers of use for non commercial purposes. *ICSLP Proceedings*, 1996.

[16] L.F. Lamel, J.L. Gauvain, B. Prouts, C. Bouhier, R. Boesch. Generation and Synthesis of Broadcast Messages, *Proceedings ESCA-NATO Workshop and Applications of Speech Technology*, September 1993.

[17] Examples include Astro Blaster, Space Fury, and Star Trek: Strategic Operations Simulator.

[18] John Holmes and Wendy Holmes. *Speech Synthesis and Recognition, 2nd Edition*. CRC: 2001. ISBN 0748408568.

[19] The HMM-based Speech Synthesis System, http://hts.sp.nitech.ac.jp/

[20] Remez, R.E., Rubin, P.E., Pisoni, D.B., & Carrell, T.D. Speech perception without traditional speech cues. *Science*, 1981, 212, 947-950.

[21] http://www.w3.org/TR/speech-synthesis/#S3.1.8

[22] Blizzard Challenge http://festvox.org/blizzard

[23] The Sound of Smiling (http://www.port.ac.uk/aboutus/newsandevents/news/title,74220,en.html)

[24] Miner, Jay et al. (1991). *Amiga Hardware Reference Manual: Third Edition*. Addison-Wesley Publishing Company, Inc. ISBN 0-201-56776-8.

[25] How to configure and use Text-to-Speech in Windows XP and in Windows Vista (http://support.microsoft.com/kb/306902)

[26] An introduction to Text-To-Speech in Android (http://android-developers.blogspot.com/2009/09/introduction-to-text-to-speech-in.html)

[27] http://www.oxil.co.uk/decSpeakToMe/modResourcesLibrary/HtmlRenderer/SpeakToMe.html

[28] Smithsonian Speech Synthesis History Project (SSSHP) 1986-2002 (http://www.mindspring.com/~ssshp/ssshp_cd/ss_home.htm)

[29] gnuspeech (http://www.gnu.org/software/gnuspeech/)

[30] http://www.abc2mp3.com

[31] http://www.acapela-group.com

[32] http://www.cereproc.com

[33] http://www.ivona.com

[34] http://www.loquendo.com

[35] http://www.neospeech.com

[36] http://meridian-one.co.uk/Orpheus.html

[37] http://www.yakitome.com

[38] http://www.voiceonthego.com

[39] http://www.voxette.co.uk

[40] Speech Synthesis Software for Anime Announced (http://animenewsnetwork.com/news/2007-05-02/speech-synthesis-software)

[41] Code Geass Speech Synthesizer Service Offered in Japan (http://www.animenewsnetwork.com/news/2008-09-09/code-geass-voice-synthesis-service-offered-in-japan)

[42] Free(b)soft - Singing Computer (http://www.freebsoft.org/singing-computer)

[43] http://bsegonnes.free.fr/multireader/en_multireader.html

[44] text to speech and RSS to Podcast (http://www.thws.cn/articles/tag/text_to_speech)reviewed by thw (http://twitter.com/thw)

- Dennis Klatt's History of Speech Synthesis (http://www.cs.indiana.edu/rhythmsp/ASA/Contents.html)

External links

- Speech synthesis (http://www.dmoz.org/Computers/Speech_Technology/Speech_Synthesis//) at the Open Directory Project

Nico Nico Douga

URL	http://www.nicovideo.jp/
Commercial?	Yes
Type of site	Video hosting service
Registration	Yes
Available language(s)	Japanese, Traditional Chinese, German and Spanish
Owner	Niwango
Created by	Niwango
Launched	December 12, 2006
Current status	Active

Nico Nico Douga (ニコニコ動画 *Niko Niko Dōga*, lit. "Smiley videos") is a popular video sharing website in Japan managed by Niwango, a subsidiary of Dwango.[1] Its nickname is "Niconico" or "Nico-dō", where "nikoniko" is the Japanese ideophone for smiling. Nico Nico Douga is the thirteenth most visited website in Japan.[2] The site won the Japanese Good Design Award in 2007,[3] and an Honorary Mention of the Digital Communities category at Prix Ars Electronica 2008.[4]

Features

Users can upload, view and share video clips. Unlike other video sharing sites, however, comments are overlaid directly onto the video, synced to a specific playback time. This allows comments to respond directly to events occurring in the video, in sync with the viewer - creating a sense of a shared watching experience. Together with Nishimura Hiroyuki serving as director at Niwango, Nico Nico Douga's atmosphere and cultural context is close to 2channel's or Futaba Channel's, and many popular videos have otaku tastes, such as anime, computer games and pop music. Nico Nico Douga offers tagging of videos. Tags may be edited by any user, not just the uploader. Each video may have up to ten tags, of which up to five may be optionally locked by the uploader, but all others may be edited by any user. Frequently these tags are used not only as categorization, but also as critical commentary, satire, or other humor related to the video's content. The site is also known for its MAD Movies and its medleys of popular songs on the website, most notably Kumikyoku Nico Nico Douga.

Other features include:

- **High video quality**: Nico Nico Douga encourages users to pre-encode their videos in a format suitable for unmodified distribution. As of July 5, 2008, H.264 video and AAC audio is supported for both free and premium users.

- **Mylist**: Each user may create 'mylists', which function similarly to a list of bookmarks. All users can create up to twenty-five mylist folders, while a basic account can have 100 videos recorded and a premium (paid) account 500 videos per mylist, giving them a total of 12,500 mylist spots. Daily mylist activity is used to compute the default ranking view, although one may also sort by view or comment count. Mylists may be optionally made public and linked to; for example, to make a list of one's own works.

- **Uploader comments**: The uploader of a video may attach permanent comments to the video. These are often used for such things as subtitles, lyrics, or corrections.

- **Nicoscript**: By using special commands in the uploader comments, the uploader can add special effects to the video, including voting, automatic transfer to another video, quiz scoring, and other features.

History

The first version of Nico Nico Douga used YouTube as a video source. However, as the site became more popular, so much traffic was transferred from YouTube that YouTube blocked access from Nico Nico Douga. Consequently Nico Nico Douga was forced to shutdown the service but two weeks later it commenced its service with its own video server. On May 7, 2007, the Nico Nico Douga for mobile phone users was announced. Since August 9, 2007, "Nico Nico Douga (RC) Mobile" has serviced mobile phones of NTT DoCoMo and au.[5]

As of September 19, 2009, Nico Nico Douga has over 14,000,000 registered users and 500,000 premium users.[6] Due to the limited server capacity, Niwango limits the amount of free users accessible to the website at peak times (7pm to 2am), based on the time of registration. The website is written in Japanese and the majority of the site traffic is from Japan, although approximately four percent is from outside of Japan, notably one percent from Taiwan.[7] A Taiwanese version of the site was launched on October 18, 2007.[8] In July 2008, the website was localized to German and Spanish.[9]

Business aspects

The main income of Nico Nico Douga is divided into three parts: Premium-Membership (Pay-membership), Advertisement, and Nico Nico Ichiba (Affiliate).[10] [11] [12]

Premium-Membership

Registration is needed to watch videos at Nico Nico Douga. There are two types of registered accounts, Free membership and Premium-membership. The Premium-membership fee is around 500 yen (about US$5) a month. As of December 12, 2009, there are 600,000 premium members.

Advertisement

Nico Nico Douga uses Google Ads and other web advertisements. On May 8, 2008, Niwango announced partnership with Yahoo! Japan, and plans to adopt search-related ads and other Yahoo-related services.

Nico Nico Ichiba (Affiliate)

Nico Nico Ichiba is a unique advertisement system in which users can place banners freely in each video page. Both video uploader and video viewer can choose items which they want to place, and can place and delete banners in the advertisement area. Users also can know how many each banners have been clicked, how many items have been bought. Ranking info of numbers of items bought through Nico Nico Ichiba is also officially provided. Items available are from Amazon.co.jp, Yahoo Shopping, and Dwango mobile service.

Premium accounts and the affiliate system are currently only available to Japanese users. Non-Japanese users can apply for a premium membership on the Japanese site with international cards, language barrier notwithstanding.

Current financial condition

As of May 9, 2008, Nico Nico Douga has had a gross income of approximately 170 million yen (US$1.42 million as of May 29, 2008), where 100 million yen (US$948,000) of that comes from paid premium memberships, 20 million yen (US$189,760) from their affiliates program, and another 50 million yen (US$474,400) from advertisements. However, operating the service costs approximately 250 million yen (US$2.37 million).

Copyright problems

Nico Nico Douga has a copyright infringement problem. On October 30, 2007, Niwango and the JASRAC, Japanese copyright holders' society agreed to form a comprehensive partnership and Niwango will pay two percent of its earnings to JASRAC as copyright royalties.[13] On March 11, 2008, Dwango announced that they would tighten up on the deletion of videos and monitoring uploaded videos. In the same way, on July 2, 2008, Dwango announced to three organizations that they would strengthen the deletion of anime and related content, including videos that use elements of copyrighted anime (known as MAD Movies). At the same time, Nico Nico Douga changed its system to expose what right holder deleted which video.

See also

- Kumikyoku Nico Nico Douga

External links

- Nico Nico Douga's official website [14] (**Japanese**)
- Smilevideo [15] (**Japanese**)
- Nico Nico Douga wiki [16] (**Japanese**)
- Nico Nico Ranking wiki [17] (**Japanese**)
- Nico Nico Pedia [18] (**Japanese**)
- Interview with [[Hiroyuki Nishimura [19]], Director at Niwango, October 27, 2007, ITmedia] (**Japanese**)
- Takeshi Natsuno, Advisor of Dwango and [[Hiroyuki Nishimura|Hiroyuki [20]] shows up at a summer Nico Nico event, July 7, 2008, ITmedia] (**Japanese**)

References

[1] "Dwango Co., Ltd. - Subsidiaries" (http://info.dwango.co.jp/english/etc/group.html). Dwango. . Retrieved 2010-01-05.
[2] "Alexa Traffic ranking" (http://www.alexa.com/topsites/countries/JP). Alexa Internet. . Retrieved 2010-01-12.
[3] "Good Design Award No.07C02037" (http://www.g-mark.org/search/Detail?id=33883&sheet=outline) (in Japanese). . Retrieved 2008-07-02.
[4] "Ars Electronica Prix Honorary Mentions" (http://www.aec.at/en/prix/winners_honorary.asp). Prix Ars Electronica. . Retrieved 2008-07-02.
[5] "Nico Nico Douga Mobile Tester Starting" (http://blog.nicovideo.jp/cat7/) (in Japanese). Nico Nico Douga Developer's Blog. 2007-05-07. . Retrieved 2008-07-02.
[6] "Over Ten Million Registered Users Notification" (http://info.niwango.jp/pdf/press/2008/20081113.pdf) (in Japanese). Niwango press release. 2008-11-13. . Retrieved 2008-11-13.
[7] "Gudadada Notice" (http://blog.nicovideo.jp/2007/08/post_148.php) (in Japanese). Nico Nico Douga Developer's Blog. 2007-08-08. . Retrieved 2007-10-12.
[8] "Nico Nico Douga's Expansion" (http://japan.cnet.com/news/media/story/0,2000056023,20358356,00.htm) (in Japanese). CNET Japan. 2007-10-10. . Retrieved 2007-10-12.
[9] "Nico Nico Douga Announcement: Specific User Function Addition" (http://www.itmedia.co.jp/news/articles/0807/04/news130.html) (in Japanese). IT Media. 2008-07-04. . Retrieved 2008-07-06.
[10] "Nico Nico Ichiba's Proceeds" (http://ascii.jp/elem/000/000/131/131583/) (in Japanese). ASCII. 2008-05-09. . Retrieved 2008-07-02.
[11] "Midway to September 2008 Balance Explanation" (http://info.dwango.co.jp/pdf/ir/news/2008/080509_ir.pdf) (in Japanese). DWANGO. 2008-05-09. . Retrieved 2008-07-02.

[12] "Yahoo! Japan and Nico Nico Douga Cooperation Start" (http://bb.watch.impress.co.jp/cda/news/21831.html) (in Japanese).
      2008-05-09. . Retrieved 2008-07-02.
[13] "Nico Nico Douga and YouTube Copyright Fee Payment" (http://www.itmedia.co.jp/news/articles/0710/30/news065.html) (in
      Japanese). IT Media. 2007-10-30. . Retrieved 2008-07-02.
[14] http://www.nicovideo.jp/index.php
[15] http://www.smilevideo.jp/index.php
[16] http://nicowiki.com/
[17] http://www39.atwiki.jp/niconicoranking/
[18] http://dic.nicovideo.jp/
[19] http://www.itmedia.co.jp/news/articles/0710/26/news032.html
[20] http://www.itmedia.co.jp/news/articles/0807/07/news024.html

Loituma Girl

Loituma Girl (also known as **'Leekspin'**) is a Flash animation set to a scat singing section of the traditional Finnish folk song "Ieva's Polka" sung by the Finnish quartet Loituma. The song is taken from the band's 1995 debut album *Things of Beauty*.[1] It appeared on the Internet in late April 2006 and quickly became popular.[2] The animation consists of a 5-frame animation of the *Bleach* anime character Orihime Inoue twirling a leek to a 27-second loop from the song. The animation loops continuously.

The "Loituma Girl" Orihime Inoue twirling her leek is the background of the Flash animation.

Content

The animation of Loituma Girl is taken from episode two of the *Bleach* anime series, between the 12 and 14 minute depending on the version. In the clip, Orihime is twirling a spring onion while talking to other characters. The scene is an instance of a recurring joke surrounding her character, in which she wants to cook something so unusual that it seems almost inedible.

The music used consists of the second half of the fifth stanza (four lines) and the complete sixth stanza (eight lines) from the song. Unlike the rest of the song, these two stanzas have no meaning, consisting mostly of phonetically-inspired Finnish words that vary from performance to performance and are usually made up on the spot by the singer (compare scat singing in jazz). These stanzas are therefore not generally listed on lyrics pages, causing confusion for people looking for lyrics that match the animation.(*See Ievan Polkka*.) There has been some confusion concerning the exact nature of the vegetable in the animation. In the Japanese version of the anime, it is identified as a Welsh onion, but the American dub identifies it as a leek, from which the name of the animation is derived. This confusion may be because Wales identifies the leek as a national symbol. However, the two are not linked.

Popularity

On 10 July 2006, the Finnish newspaper *Helsingin Sanomat* reported that Loituma Girl had caused a resurgence in Loituma's popularity, and the band had received thousands of fan letters from around the world.[3]

The animation quickly became popular amongst hundreds of thousands of World of Warcraft fans who were spreading the links to various websites containing the animated video via forums, chat programs, voice communication channels and in-game chat.

Band member Timo Väänänen describes his initial reaction to the video:[2]

> I first found out there was something going on when I looked at the statistics of my own web page and then I realized that something weird is happening because there is such a huge traffic there. And most of the traffic came from Russia and then I started to track down what is happening and then I found this video. And well, I have no idea what this video is about, and what this girl is about.

PRI's *The World* radio program even covered the animation in a segment, in which they noted the clip's trance-inducing qualities. Patrick Macias, who was interviewed in the program, described the animation:[2]

> This is basically a joke for someone who spends all of their time staring at a computer, made by people who spend all of their time staring at a computer. It's possible to read deeper meanings into it, but it sort of defeats the purpose because in the end it's just this hypnotic clip of animation.

As with most Internet phenomena, there are numerous videos, remixes, and parodies that have been inspired by the Flash animation. These may feature the animated background, the song clip, or otherwise reference the style of the animation.

Commercialization

In August 2006, German ringtone provider Jamba! began selling a collection of media based on the animation. The video shows an anthropomorphic donkey (called *Holly Dolly*) dancing to the animation which is displayed (flipped horizontally) in the background.[4] The song/animation is marketed as the "Dolly Song", and the music is played faster than the original Loituma version. It was also given an extra 30-second drum preface, which was not present in the original version.

In January 2007, a similar video, entitled "Holly Dolly - Dolly Song (Ieva's Polka)", appeared in the Google Video Top 100, though it was present on the Internet for a while before. It features the same donkey, along with some dancing sheep and a snowman, but the leek-spinning girl in the background is only there briefly.[5] In April 2007, a Dutch power company (Eneco) used the song/melody of Loituma Girl in its TV commercial for "ecostroom" (green energy).[6]

In May 2007, Wrigley's used this song in their German TV spot for their Extra gum.[7] The Dutch company Artiq Mobile launched a website where people can upload home-made Loituma girl spoof videos. TV commercials state the best video will win 500 euros.[8] The Romanian company Romtelecom uses the song in one of their commercials for Dolce, a satellite television service.

In October 2007, McDonald's Hungary used this song in their Hungarian TV spot for their McCafé commercial.[9]

In early 2009, Ready Brek (UK) used a gibberish version of the song in their advertisement for cereal.

Many YouTubers animate their own versions of Leekspin with different characters and spinning objects.

Vocaloid adaption

In September 2007 Yamaha's Vocaloid voice synthesizer was given a makeover with the Character Vocal Series, featuring an anime-style character named Hatsune Miku as the mascot for the software. One of the initial promotional videos set the already popular Loituma Girl song to an animation of a chibi version of Miku Hatsune waving a Welsh onion, using Vocaloid software for the voice.[10] To date, this video has over nine million views on Nico Nico Douga and YouTube combined.[11] [12]

The popularity of the Vocaloid version of Ievan Polkka has led to the chibi-Miku featured in the video being treated as a separate character, Hachune Miku, who has even gotten a Nendoroid action figure produced in her image by Good Smile Company.

See also

• Ievan Polkka
• Internet phenomenon
• Orihime Inoue
• Internet Meme

External links

• The flash animation [13]
• The Holly Dolly video [14]
• Leekspin.com [15]

References

[1] "NorthSide Catalog - Things of Beauty - Loituma" (http://www.noside.com/Catalog/CatalogAlbum_01.asp?Album_ID=45). .

[2] Werman, Marco (2006-08-18). "Global Hit" (http://www.theworld.org/?q=node/3625) (radio). *The World*. Public Radio International. . Retrieved 2006-08-18.

[3] "Hittibiisi kulman takaa" (http://www.hs.fi/uutiset/verkko-hesari/artikkeli/Hittibiisi+kulman+takaa/1135220601104). *Helsingin Sanomat*. 2006-07-10. .

[4] *Holly Dolly Song* (http://order.jamba.de/storage/view/93/do/HollyDollyDollySong2.mpg). [MPEG]. Jamba!. .

[5] http://video.google.com/videoplay?docid=2295360340367941531&hl=en

[6] "Dutch Eneco commercial" (http://youtube.com/watch?v=6brHh8FArrA). .

[7] http://www.wrigley.de/images/movies/extra_trackfinal.mp3

[8] "Preidols.nl - Jouw eigen meisje met de prei!" (http://preidols.nl/) (in Dutch). . Retrieved 2007-06-11.

[9] http://www.youtube.com/watch?v=lVCHi6GJLOE

[10] "How Hatsune Miku opened the creative mind: Interview with Crypton Future Media" (http://www.itmedia.co.jp/news/articles/0802/25/news017.html) (in Japanese). February 25, 2008. . Retrieved 2008-02-29.

[11] http://www.nicovideo.jp/watch/sm982882

[12] http://www.youtube.com/watch?v=kbbA9BhCTko

[13] http://dagobah.biz/flash/loituma.swf

[14] http://video.google.com/videoplay?docid=2295360340367941531&hl=en

[15] http://www.leekspin.com/

Cameo appearance

A **cameo role** or **cameo appearance** (often shortened to just **cameo**) is a brief appearance of a known person in a work of the performing arts, such as plays, films, video games[1] and television. Short appearances by film directors, politicians, athletes, musicians, celebrities or even characters from another fictional work are common. These roles are generally small, and many of them are non-speaking.

History

Originally the phrase "cameo role" referred to a famous person who was playing no character, but him or herself. Like a cameo brooch—a low-relief carving of a person's head or bust—the actor or celebrity is instantly recognizable. More recently, "cameo" has come to refer to any short appearances, whether as a character or as oneself.

Cameos are often noncredited due to their shortness or because of a perceived mismatch between the celebrity's stature and the film or TV show in which he or she is appearing. Many are publicity stunts. Others are acknowledgments of an actor's contribution to an earlier work, as in the case of many film adaptations of TV series, or of remakes of earlier films. Others honour artists or celebrities known for work in a particular field.

A cameo can establish a character as being important without having much screen time. Examples of such cameos are Sean Connery in *Robin Hood: Prince of Thieves*, Ted Danson in *Saving Private Ryan*, or George Clooney in *The Thin Red Line*.

Cameos are also common in novels and other literary works. "Literary cameos" usually involve an established character from another work who makes a brief appearance in order to establish a shared universe setting, to make a point, or to offer homage. Balzac was an originator of this practice in his "Comedie humaine". Sometimes a cameo features a historical person who "drops in" on fictional characters in a historical novel, as when Benjamin Franklin shares a beer with Phillipe Charbonneau in "The Bastard" by John Jakes. A cameo appearance can also be made by the author of a work in order to put a sort of personal "signature" on a story. An example from the thriller genre includes Clive Cussler, who made appearances in his own novels as a "rough old man" who advised action hero Dirk Pitt. An example in the comic book genre is John Byrne's resplendent use of cameos in Marvel Comics' "Iron Fist" #8, which features appearances by Byrne himself, Howard the Duck (on a poster), Peter Parker and Mary Jane Watson, Sam McCloud, Fu Manchu, and Wolverine.

At the apex of the technique stands "Lolita" by Vladimir Nabokov. This acclaimed novel is, among many other things, a "tour de force" of literary cameos.

Early appearances are often mistakenly considered as cameos. Sylvester Stallone appears in Woody Allen's *Bananas* credited as only as "Subway Thug #1", five years before his breakout role in 1976's *Rocky*, therefore making it an early appearance of a non-established actor.

Examples of cameos

Directors

Directors often appear in cameo roles to add a personal "signature" on a film. Alfred Hitchcock often enjoyed inserting himself, as a passive by-stander, in scenes of his films. Such cameo appearances in 37 of his movies helped popularise the term among general audiences. Often whimsical, the cameos became so well publicised that audiences began watching for them. Consequently, Hitchcock began placing the cameos early in each film so audiences could then give their full attention to the story. Director Sam Raimi has followed Hitchcock's example in many of his films-for example, he is the second student who hits Peter Parker in the head with his bookbag at Empire State University in *Spider-Man 2*. Anthony Zuiker has appeared in several cameos throughout his hugely popular *CSI:*

Crime Scene Investigation primetime television show. Terry Gilliam has appeared in "Brazil" as a randomly peculiar character in an overcoat smoking a cigarette (with a trail of cigarette butts in the hallway) upon Sam Lowry's return to his apartment. Gilliam has also appeared in Jabberwocky as a "stone miner" and in general, had similarly bizarre and brief roles in the Monty Python films, which he co-directed.

Other directors are known for casting themselves in cameo roles in their films. Quentin Tarantino provides cameos or small roles on some of his movies. M. Night Shyamalan appears in some of his movies, such as *The Village*, in which he is shown in the glass reflection of the sheriff, and also as a shady fan at a stadium in *Unbreakable*. In *The Sixth Sense* he is shown to be the doctor at the hospital and has a brief appearance in a short scene with the child's mother. In *Signs* he is the vet Ray Reddy, who is involved in the accident that took Graham's wife's life.

Likewise, Peter Jackson has made brief cameos in all of his movies, except for the puppet movie *Meet the Feebles*. For example, he plays a peasant eating a carrot in *The Fellowship of the Ring*; a Rohan warrior in *The Two Towers* and a pirate boatswain in *The Return of the King*. All three were non-speaking "blink and you miss him" appearances. He also appears in his 2005 remake of *King Kong* as the gunner on a biplane in the finale.

Director Martin Scorsese appears in the background of his films as a bystander or an unseen character. In *Who's That Knocking at My Door*, he appears as one of the gangsters, a passenger in *Taxi Driver*. He opens up his 1986 film *The Color of Money* with a monologue on the art of playing pool. In addition, he appears with his wife and daughter as wealthy New Yorkers in *Gangs of New York*, and he appears as a theatre-goer and is heard as a movie projectionist in *The Aviator*.

Actors and writers

In the film version of Hunter S Thompson's book Fear and Loathing in Las Vegas starring Johnny Depp as Raoul Duke, Hunter S Thompson's alter-ego, Thompson can be seen quickly as an older version of Depp's character in a flashback scene at a San Francisco nightclub. Similarly, Arthur C. Clarke makes a brief cameo appearance in the film adaptation of his book *2010: Odyssey Two*. S.E. Hinton played a nurse in the film adaptation of her novel, *The Outsiders*. Stephenie Meyer appears eating at a diner in the film adaptation of her novel, *Twilight*. In the 2009 film *The Invention of Lying*, there were cameos from Edward Norton as a cop, Philip Seymour Hoffman as a bartender, Christopher Guest, and Stephen Merchant. In the recent film adaptation of author, Sapphire's, 1996 novel, Push, (renamed, *Precious*, so not to be confused with the 2009 action film of the same name), Sapphire appears in one of the end scenes as the woman running the daycare.

Remakes and sequels occasionally feature actors from the original films. In the 2003 version of "Willard" the framed picture of Willard's father is a picture of Bruce Davison, who played Willard in the 1971 version of the film. The 2004 version of *Dawn of the Dead* features cameos by Ken Foree, and Scott Reiniger. The original stars of Starsky and Hutch appeared at the end of the 2004 film, and Bernie Kopell, who portrayed Siegfried in the original show appeared in the 2008 film version of Get Smart. Vin Diesel made a short appearance at the end of *The Fast and the Furious: Tokyo Drift* where he challenges to race Shawn, Lucas Black, the then Drift king. The 2005 remake of "The Longest Yard" features Burt Reynolds (as the coach, Nate Scarboro, who was previously played by Michael Conrad), who starred as Paul Crewe in the in original 1974 film. However, his role is not considered to be a cameo due to him being one of the lead actors.

In the same vein as the remake and sequel, actors can also make appearances in completely different films which are directed by or star another actor they are friendly with. Actors Ben Stiller, Vince Vaughn, Owen Wilson, Luke Wilson, and Will Ferrell and others made appearances in so many of the same films (whether as lead characters or cameos) *USA Today* coined the term the "Frat Pack" to name the group.[2] Actor Adam Sandler is also known for frequently casting fellow Saturday Night Live performers (including Rob Schneider and David Spade) in various roles in his films (as well as making cameo appearances of his own in theirs, most of which he co-produces). Sam Raimi frequently uses his brother Ted and Bruce Campbell in his films. [3] [4]

Directors can also be known to cast well-known lead actors which they have worked with in the past in cameo roles for other films. Among the many cameos featured in the film *Maverick*, (directed by Richard Donner), actor Danny Glover (Mel Gibson's co-star in the *Lethal Weapon* franchise of films also directed by Donner) appears as the lead bank robber. He and Maverick (Gibson) share a scene where they look as if they knew each other, but then shake it off. As Glover makes his escape with the money, he mutters "I'm too old for this shit.", his character's catch phrase in all four *Lethal Weapon* films. In addition, a strain of the main theme from *Lethal Weapon* plays in the score when Glover is revealed. Actress Margot Kidder made a cameo appearance in the same film as a robbed villager. Kidder starred as Lois Lane in Superman, also directed by Donner.

Real life people

Films based on actual events occasionally include cameo roles of the people portrayed in them. In the 2006 film *The Pursuit of Happyness*, Chris Gardner makes a cameo in the end. *24 Hour Party People*, a film about Tony Wilson has a cameo by the real Tony Wilson. In the film *Apollo 13*, James Lovell (the real commander of that flight) appeared at the end, shaking hands with Tom Hanks. Domino Harvey makes a short appearance in the credits of *Domino*. The real Erin Brockovich has a cameo appearance as a waitress named Julia in the movie named after herself (where her role is played by actress Julia Roberts). The 2000 film *Almost Famous* featured *Rolling Stone* co-founder Jann Wenner as a passenger in a New York City taxicab. Chuck Yeager has a cameo as "Fred," a bartender at "Pancho's Place", in *The Right Stuff*. In the 2008 film *21*, Jeff Ma, the character the film is based on, plays a blackjack dealer at the Planet Hollywood Resort and Casino. His character in the movie calls him "my brother from another mother".

In a similar vein, cameos sometimes feature persons noted for accomplishments outside the film industry, usually in ways related to the subject or setting of the film. *October Sky* (1999), set in 1950s Appalachia, featured photographer O. Winston Link in a brief appearance portraying a steam locomotive engineer. Link became famous in the 1950s for chronicling the last days of regular steam locomotives service in the region. *O Brother, Where Art Thou?* (2000), set in Depression-era rural South, featured cameos by country "roots" music notables such as Alison Krauss, Ralph Stanley, Gillian Welch, The Whites and the Fairfield Four. In the film *The Last Mimzy*, noted string theorist Brian Greene has a cameo as the Intel scientist. In Dr. Dolittle 2 a cameo appearance was made by Steve Irwin. Stan Lee, the creator of many Marvel Comics characters has appeared in the film versions of the comics, including X-Men, Spider-Man, Iron Man,The Incredible Hulk and The Fantastic Four. Skateboarder Tony Hawk makes a cameo as a dead body in an episode of *CSI Miami*. In *The Fast and the Furious: Tokyo Drift* during the first scene, Keiichi Tsuchiya, the professional drifter, makes an appearance as a fisherman. On the plane that Shawn takes to Japan, the seat in front of him is occupied by Rhys Millen, a stunt driver (who also did many of the stunts in the movie). In *Men in Black 2*, Biz Markie (a hip hop artist) appears as an alien who uses beatboxing to communicate.

Mike Todd's film *Around the World in Eighty Days* (1956) was filled with cameo roles: (John Gielgud as an English butler, Frank Sinatra playing piano in a saloon), and others. The stars in cameo roles were pictured in oval insets in posters for the film, and gave the term wide circulation outside the theatrical profession. Notably the *1983 television adaptation* and *2004 film version* of the story also feature a large number of cameos.

It's A Mad, Mad, Mad, Mad World (1963), an "epic comedy", also features cameos from nearly every popular American comedian alive at the time, including a silent appearance by the Three Stooges and a voice-only cameo by Selma Diamond. In the apolatic zombie movie Zombieland, the main characters go to Bill Murray's mansion, wheras he lives.

·

Fictional characters

A type of fictional crossover is the placement of two or more otherwise discrete fictional characters into the context of a single story. This occurrence can arise from legal agreements between the relevant copyright holders, or because of unauthorized efforts by fans and is intended for promotional, parodic or other purposes. John Munch, a fictional detective played by actor Richard Belzer, which first appeared on *Homicide: Life on the Street*, made, among numerous other TV show crossovers, a small cameo appearance in the episode "Took" from the fifth and final season of *The Wire*. In *Homicide*, along with Tim Bayliss (played by Kyle Secor) and Meldrick Lewis (played by Clark Johnson), Munch is co-owner of "The Waterfront", a bar located across the street from their Baltimore police station. In *The Wire* he refers to owning "The Waterfront" in the past-tense and talks about wanting to buy a bar again in New York City in the crumbling economy of the country.[5]

See also

- Bit part
- Extra
- Self-insertion
- List of Hitchcock cameo appearances
- List of directors who appear in their own films
- Crossover fiction

References

[1] Michael Donahue, "Forced Guests: Cameos that make us sceam 'Yessss!'" in *Electronic Gaming Monthly* 226 (March 2008): 34.

[2] Wloszczyna, Susan (2004-06-15). "Wilson and Vaughn: Leaders of the 'Frat Pack'" (http://www.usatoday.com/life/movies/news/2004-06-15-frat-pack_x.htm). *USA Today*. .

[3] http://www.monsters-movies.com/sam_raimi.htm

[4] http://www.imdb.com/name/nm0000600/workedwith

[5] Rob Owen (2008). "Tuned In Journal: Munch on *The Wire*" (http://www.post-gazette.com/pg/08043/856709-237.stm). Pittsburgh Post-Gazette. . Retrieved 2010-01-07.

Victor Entertainment

&Victor · JVC	
Type	Subsidiary of Japan Victor Company (JVC)
Founded	Yokohama, Japan (April 25,1972)
Headquarters	Kita-Aoyama, Minato-ku, Tokyo, Japan
Key people	Yuichi Kaito, President Toshiaki Shibutani, Managing Director
Industry	Entertainment
Products	Audio, visual and computer software media products
Revenue	6.31 billion Yen (Fiscal year ended March 31, 2005)
Employees	600 (as of March 31, 2005)
Website	Victor Entertainment [1]

Victor Entertainment (ビクターエンタテインメント株式会社 *Bikutā Entateinmento Kabushiki Kaisha*) is a subsidiary of Japan Victor Company (JVC) that produces and distributes music, movies and other entertainment products such as anime and television shows in Japan. It was formerly known as **Victor Music Industries** (ビクター音□産業 *Bikutā Ongaku Kangyō*).

History

- **April 1972** — Victor Music Industries, Inc. (ビクター音□産業株式会社 *Bikutā Ongaku Sangyō Kabushiki Kaisha*) spun off as subsidiary of JVC
- **February 1984** — Sales and marketing department of JVC spun off as Nihon AVC, Inc. (日本エイ・ブイ・シー株式会社 *Nihon Ei · Bui · Shī Kabushiki Kaisha*)
- **October 1987** — Victor Music Industries striked a deal with BMG Japan after RCA Records sold its interests to BMG
- **April 1993** — Nihon AVC and Victor Music Industries merge and the name is changed to Victor Entertainment
- **March 1999** — Moved main office
- **October 1999** — Transferred deal to BMG Funhouse
- **October 2005** — Transferred deal back to BMG Japan
- **January 2009** — Sony Music Entertainment Japan bought BMG Japan

Labels

Records

- 3 Views
- Aosis Records
- BabeStar
- Cypress Showers
- Globe Roots
- Happy House
- Hihirecords (for babies and kids)
- Invitation
- JVC Entertainment (Production and Artist Managements)

  - Flying Dog (Animation Related Products)
- JVC Jazz
- JVC World Sounds
- Mob Squad
- Nafin
- Speedstar International
- Speedstar Records
- Taishita (Southern All Stars Private Label)
- Victor
- Victory (defunct; not same as US label)

Distribution

- Bad News
- Daipro-X
- Marquee, Inc.
- Substance
- Teichiku Entertainment (Subsidiary of Victor Entertainment.)

  - BAIDIS
  - Be-tam-ing
  - Continental
  - Imperial Records
  - Imperial International
  - KIDSDOM (Animation Related Label)
  - MONAD (Haruomi Hosono Private Label)
  - Non-Standard(Haruomi Hosono Produce Label)
  - Overseas Records
  - PROGRAM (Katsumi Tanaka Private Label ~2000)
  - TOHO Records (Master Rights Only)
  - Takumi Note
  - Teichiku Records
  - TMC Music

    - 246 Records
  - Union Records

    - Union Black Records

Video

- 20th Century Fox Home Entertainment
- GAGA Communications
- MediaNet Pictures
- SPO Entertainment
- TBS Video

Major artists

Listed alphabetically by group or family name. Names are in Western order (given name, family name).

- AKINO from bless4 (flying DOG)
- Anthem
- Alesha
- ALI PROJECT (flying DOG)
- Akino Arai (flying DOG)
- Angra
- Mika Arisaka
- The Back Horn
- Blessed By A Broken Heart
- Steve Barakatt
- Chocolate & Akito
- Cocco
- The Cobra Sisters
- cymbals
- Daigo☐Stardust
- Death from Above 1979 (Japanese distribution rights)
- Leah Dizon
- Dragon Ash
- FictionJunction (flying DOG)
  - FictionJunction YUUKA (flying DOG)
- Full of Harmony
- Fūmidō
- Gari
- Going Under Ground
- Guniw Tools
- 80☐PAN!
- Impellitteri
- Jero
- Kazuyoshi Saito
- Miho Hatori
- Chiaki Ishikawa (flying DOG)
- Yoko Kanno (flying DOG)
- Yuki Kajiura (flying DOG)
- Kigurumi
  - Kei (joined "Kigurumi" on November 7, 2007.)
  - Miki (joined "Kigurumi" on November 7, 2007.)
  - Rena

- Kigurumichiko
  - Rena (Kigurumi)
  - Michiko Shimizu
- Kiko Loureiro
- Kiroro
- Kokia
- Kyōko Koizumi
- Lisa Komine (flying DOG)
- Love Psychedelico
- Lunkhead
- Masumi
- Matt Bianco
- May'n (flying DOG)
- Merry
- Minmi
- Miz
- The Morning After
- Megumi Nakajima (flying DOG)
- Takeshi Nakatsuka
- Rimi Natsukawa
- Aina Ōgi
- Kiyofumi Ōno
- Oozekitaku
- Paris Match
- PIG
- Quruli
- Remioromen
- ROUND TABLE featuring Nino (flying DOG)
- Sakanaction
- Shinichi Mori
- Noriko Sakai
- Maaya Sakamoto (flying DOG)
- savage genius (flying DOG)
- See-Saw (flying DOG)
- Shunsuke Kiyokiba (former EXILE vocalist that was signed under Avex)
- Shōnen Kamikaze
- Singer Songer
- SMAP
  - Goro Inagaki (also goes by &G)
  - Shingo Katori
  - Tsuyoshi Kusanagi
  - Masahiro Nakai
- Source
- Southern All Stars
- Sweet Vacation
  - Yuko Hara (solo and duet releases with SMAP member Shingo Katori)

- Keisuke Kuwata/Kuwata Band
- Hiroshi Matsuda
- Hideyuki Nozawa
- Kazuyuki Sekiguchi
- Toshihiko Tahara (transferred from Pony Canyon)(Management Only)
- Mariko Takahashi
- Tokyo Ethmusica
- UA
- Aimi Yuguchi
- Yukio Hashi
- Sayuri Yoshinaga

Distribution managements

- BMG Japan (1987-1999), (2005-2008)
- BMG Funhouse (1999-2005)
- Sony Music Entertainment Japan (2009-present)

Note that Victor Music Industries was self distributing from 1972 to 1987.

External links

- **(Japanese)** Victor Entertainment [2]

References

[1] http://www.jvcmusic.co.jp/
[2] http://www.jvcmusic.co.jp

Animelo Summer Live

Animelo Summer Live	
Location(s)	Japan
Years active	2005-present
Date(s)	Summer
Genre	Anime music
Website	Animelo Summer Live 2009 -RE:BRIDGE- [1]

Animelo Summer Live is the biggest annual anime songs concert in Japan managed by *Animelo Mix*. Animelo Summer Live has been held every summer since it was first held in 2005.

Description

Animelo performers are singers and seiyuu that specialize in singing anime or game theme songs. The performers are not necessarily under the same record label. Animelo's performers are often called *Anisama Friends*, and are represented by famous anisong singer and Animelo producer Masami Okui. The highlights of Animelo concerts are the collaboration between the performers.

Animelo has a different theme song for each year, which is sung by all performers in the end of the concert.

2005

- Title: **Animelo Summer Live 2005 -THE BRIDGE-**
- Date: July 10, 2005
- Location: Yoyogi National Gymnasium
- Sponsors: Dwango, QR
- Support: FM NACK5
- Collaboration: evolution, Giza Studio, King Records, Geneon, Bellwood Records, Pony Canyon, Lantis, Realize Records

Performers

- Masami Okui
- Hironobu Kageyama
- JAM Project
- Nana Mizuki
- Naozumi Takahashi
- Minami Kuribayashi
- Chihiro Yonekura
- Yoko Ishida
- can/goo
- Mikuni Shimokawa
- Unicorn Table
- Tatsuhisa Suzuki
- Tomoe Ohmi
- Rina Aiuchi

Set List

1. Transmigration / Nana Mizuki, Masami Okui (*ES Hour Radio Time* image song)
2. Rondo-revolution (輪舞-revolution) / Masami Okui (*Revolutionary Girl Utena* OP)
3. A Confession of TOKIO / Masami Okui (CM song for *Animelo Mix*)
4. TRUST / Masami Okui (*He Is My Master* OP)
5. Open Your Mind ~chiisana hane hirogete~ (OPEN YOUR MIND〜小さな羽根ひろげて) / Yoko Ishida (*Ah! My Goddess* OP)
6. Jounetsu no Megami (情熱の女神) / Yoko Ishida (*Anime TV* ED)
7. Zankoku na Tenshi no These (残酷な天使のテーゼ) / Yoko Ishida, Masami Okui (*Neon Genesis Evangelion* OP)
8. We Are! (ウィーアー!) / Hiroshi Kitadani, Masaaki Endoh, Naozumi Takahashi (*One Piece* OP)
9. Kujikenaikara! / Minami Kuribayashi, Mikuni Shimokawa (*Slayers* ED)
10. CHA-LA HEAD CHA-LA / Hironobu Kageyama (*Dragon Ball Z* OP)
11. Yume Kounen (夢光年) / Hironobu Kageyama (*Uchuusen Sagitarius* ED)
12. Just A Survivor / Tatsuhisa Suzuki (*Sukisho* OP)
13. tomorrow / Mikuni Shimokawa (*Full Metal Panic!* OP)
14. Minami kaze (南風) / Mikuni Shimokawa (*Full Metal Panic!: The Second Raid* OP)
15. Haruka, Kimi no Moto e... (遙か、君の◻とへ...) / Naozumi Takahashi (*Haruka: Beyond the Stream of Time* OP)
16. glorydays / Naozumi Takahashi (*S.S.D.S. ~setsuna no eiyuu~ (S.S.D.S.〜刹那の英雄〜)* image song)
17. kun ni ae te yokatta (君に会えてよかった) (iromelo-mix) / Naozumi Takahashi (*Radio Animelo Mix* theme song)
18. FLY AWAY / unicorn table (*Jinki:EXTEND* OP)
19. Arashi no Naka de Kagayaite (嵐の中で輝いて) / Chihiro Yonekura (*Mobile Suit Gundam: The 08th MS Team* OP)
20. Eien no Tobira (永遠の扉) / Chihiro Yonekura (*Mobile Suit Gundam: The 08th MS Team Miller's Report* OP)
21. Anisama Special Medley / Chihiro Yonekura
    - WILL (*Hōshin Engi* OP)
    - Little Soldier (*PowerPro* Game OP)
    - Boku no SPEED de (僕のスピードで) (*Mahoraba* ED)
22. Utakata (ウタカタ) / Tomoe Ohmi (*Dream Factory* ED)
23. Maboroshi (まぼろし) / can/goo (*Sister Princess: RePure* OP)
24. Oshiete ageru (教えてあげる) / can/goo (*Doki Doki School Hours* OP)
25. Precious Memories / Minami Kuribayashi (*Kimi ga Nozomu Eien* OP)
26. Blue Treasure / Minami Kuribayashi (*Tide-Line Blue* OP)
27. Muv-Luv (マブラヴ) / Minami Kuribayashi (*Muv-Luv* OP)
28. SKILL / JAM Project (*2nd Super Robot Wars Alpha* OP)
29. Genkai Battle (限界バトル) / JAM Project (*Yu-Gi-Oh! GX* ED)
30. Meikyuu no Prisoner (迷宮のプリズナー) / JAM Project (*Super Robot Wars Original Generation: The Animation* OP)
31. VICTORY / JAM Project, Yoko Ishida, Chihiro Yonekura (*Super Robot Wars MX* OP)
32. I can't stop my love for you / Rina Aiuchi (*Detective Conan* OP)
33. Koi Wa, Thrill, Shock, Suspense (恋はスリル、ショック、サスペンス) / Rina Aiuchi (*Detective Conan* OP)
34. Still in the groove / Nana Mizuki (*Iromelo Mix* CM song)
35. Take a shot / Nana Mizuki (*Magical Girl Lyrical Nanoha* insert song)
36. WILD EYES / Nana Mizuki (*Basilisk* ED)
37. POWER GATE / Female performers

38. ONENESS / Anisama Friends

    -encore-

39. ACCESS! / Naozumi Takahashi, Yoko Ishida, Masaaki Endoh, Masami Okui, Minami Kuribayashi, Mikuni Shimokawa, Tatsuhisa Suzuki, Nana Mizuki, Chihiro Yonekura (*Radio Animelo Mix* theme song)

40. ONENESS / All performers

Media

CD

ONENESS

   Release date: June 24, 2005

   Lyrics, composition: Masami Okui

   Animelo Summer Live 2005 -THE BRIDGE- theme song

2006

- Title: **Animelo Summer Live 2006 -OUTRIDE-**
- Date: July 8, 2006
- Location: Nippon Budokan
- Sponsors: Dwango, QR
- Support: Kids Station
- Collaboration: Avex Trax, evolution, On The Run, Giza Studio, King Records, Geneon, Victor Entertainment, Bellwood Records, Lantis, Realize Records

Performers

- Masami Okui
- JAM Project
- Nana Mizuki
- Naozumi Takahashi
- Minami Kuribayashi
- Chihiro Yonekura
- Yoko Ishida
- Rina Aiuchi
- Ali Project
- U-ka saegusa IN db
- Chiaki Ishikawa
- savage genius
- KENN with The NaB's
- Aiko Kayo

Special Guest

- Aya Hirano, Minori Chihara, Yūko Gotō (as The Melancholy of Haruhi Suzumiya team)

Set List

1. MASK / Masami Okui, Minami Kuribayashi (*Sorcerer Hunters* ED)
2. Rumbling hearts / Minami Kuribayashi (*Kimi ga Nozomu Eien game* OP)
3. Crystal Energy / Minami Kuribayashi (*My-Otome* OP)
4. Shiawase no Iro (幸せのいろ) / Yoko Ishida (*Ah! My Goddess: Everyone Has Wings* OP)
5. Aka no Seijaku (紅の静寂) / Yoko Ishida (*Shakugan no Shana* ED)
6. Denkousekka no koi (電光石火の恋) / Naozumi Takahashi (*Haruka: Beyond the Stream of Time* image song)
7. Muteki na Smile (無敵な smile) / Naozumi Takahashi (*Muteki Kanban Musume* ED)
8. OK! / Naozumi Takahashi
9. Female performers medley

   - arashi no naka de kagayaite (嵐の中で輝いて) / Chihiro Yonekura (*Gundam 08th MS Team* OP)
   - Otome no Policy (乙女のポリシー) / Yoko Ishida (*Sailor Moon R* OP)
   - Shining Days / Minami Kuribayashi (*My-HiME* OP)
   - Mezase Pokémon Master (めざせポケモンマスター) / Rica Matsumoto (*Pokémon* OP)
10. Hare Hare Yukai (ハレ晴レユカイ) / Aya Hirano, Minori Chihara, Yūko Gotō (The Melancholy of Haruhi Suzumiya ED)
11. Wake Up Your Heart / KENN with The NaB's (*Yu-Gi-Oh! GX* ED)
12. Forever... / savage genius (*Elemental Gelade* OP)
13. Inori no Uta (祈りの詩) / savage genius (*Simoun* ED)
14. Aishitene motto (愛してね♥っと) / Aiko Kayo (*Tenjho Tenge* ED)
15. Hitomi no Naka no Meikyū (瞳の中の迷宮) / Aiko Kayo (*Yami to Bōshi to Hon no Tabibito* OP)
16. WILL / Chihiro Yonekura (*Hoshin Engi* OP)
17. Aozora to kimi e (青空とキミへ) / Chihiro Yonekura
18. Male performers medley

   - CHA-LA HEAD CHA-LA / Hironobu Kageyama
   - PLANET DANCE / Yoshiki Fukuyama (*Macross 7* song)
   - We Are! (ウィーアー!) / Hiroshi Kitadani (*One piece* OP)
   - Yuusha-Oh Tanjou! (勇者王誕生！) / Masaaki Endoh (*GaoGaiGar* OP)
   - Soldier Dream ~Saint Shinwa~ (SOLDIER DREAM～聖士神話～) / Hironobu Kageyama, Yoshiki Fukuyama, Hiroshi Kitadani, Masaaki Endoh (*Saint Seiya* OP)
19. Candy Lie / r.o.r/s (Masami Okui, Chihiro Yonekura)
20. SECOND IMPACT / Masami Okui (*Animelo Mix* CM song)
21. WILD SPICE / Masami Okui (*Muteki Kanban Musume* OP)
22. zero-G- / Masami Okui (*Ray the Animation* OP)
23. Kimi to Yakusoku Shita Yasashii Ano Basho Made (君と約束した優しいあの場所まで) / U-ka Saegusa in dB (*Detective Conan* OP)
24. Everybody Jump / U-ka Saegusa in dB (*Fighting Beauty Wulong Rebirth* OP)
25. 100 Mono Tobira (100 の扉) / Rina Aiuchi, U-ka Saegusa (Chorus: Sparkling Point) (*Detective Conan* OP)
26. GLORIOUS / Rina Aiuchi (*Another Century's Episode 2* image song)
27. MIRACLE -Allegro. vivacemix- / Rina Aiuchi (*MÄR* ED)
28. Seishoujo Ryouiki (聖少女領域) / ALI PROJECT (*Rozen Maiden* OP)
29. Bōkoku Kakusei Catharsis (亡國覚醒カタルシス) / ALI PROJECT (*.hack//Roots* ED)
30. GONG (Album version) / JAM Project (*Super Robot Wars Alpha 3* OP)

31. Garo ~Savior in the Dark~ (牙狼〜SAVIOR IN THE DARK〜) / JAM Project (*GARO* OP)
32. SKILL / JAM Project, Chihiro Yonekura, Minami Kuribayashi, Yoko Ishida
33. Utsukushikereba sore de ii (美しければそれでいい) / Chiaki Ishikawa (*Simoun* OP)
34. Anna ni Issho datta no ni (あんなに一緒だったのに) / Chiaki Ishikawa (*Mobile Suit Gundam SEED* ED)
35. Super Generation / Nana Mizuki (*Yaguchi Hitori* ED)
36. Innocent Starter / Nana Mizuki (*Magical Girl Lyrical Nanoha* OP)
37. Hime Murasaki (ヒメムラサキ) / Nana Mizuki (*Basilisk* ED)
38. Eternal Blaze / Nana Mizuki (*Magical Girl Lyrical Nanoha A's* OP)
39. OUTRIDE / Anisama Friends

    -encore-
40. ONENESS / Anisama Friends
41. OUTRIDE / All performers

Media

CD

OUTRIDE

> Label: Geneon
>
> Release date: June 9, 2006
>
> Lyrics: Masami Okui
>
> Composition: Hironobu Kageyama
>
> Arrangement: Yougo Kouno
>
> Animelo Summer Live 2006 -OUTRIDE- theme song

DVD

Animelo Summer Live 2006 -OUTRIDE- I

> Label: King Records
>
> Release date: December 21, 2006

Animelo Summer Live 2006 -OUTRIDE- II

> Label: Victor Entertainment
>
> Release date: December 21, 2006

✳Hare Hare Yukai performance is not included in both DVD

2007

- Title: **Animelo Summer Live 2007 Generation-A**
- Date: July 7, 2007
- Location: Nippon Budokan
- Sponsors: Dwango, QR
- Support: Kids Station, Television Kanagawa
- Collaboration: Avex Trax, avec tune, evolution, King Records, Columbia Music Entertainment, Geneon, Sistus Records, DMP Records, FIX Records, Lantis, Realize Records

Performers

- Masami Okui
- JAM Project
- Nana Mizuki
- Naozumi Takahashi
- Minami Kuribayashi
- Ali Project
- Haruko Momoi
- Jyukai
- m.o.v.e
- Tomoe Ohmi
- Psychic Lover
- Suara
- Cy-Rim rev.
- Minori Chihara

Special Guest

- Aya Hirano
- Yabeno Hikomaru (Denchu) & Kotohime (MOMOEIKA)

Secret Guest

- Yoko Takahashi
- PaniCrew

Set List

1. Rondo-revolution (輪舞-revolution) / Masami Okui, Nana Mizuki (*Revolutionary Girl Utena* OP)
2. -w- / Masami Okui
3. Sora ni kakeru hashi (空にかける橋) / Masami Okui (*Tales of Eternia* OP)
4. Romantic summer / Halko Momoi (*Seto no Hanayome* OP)
5. WONDER MOMO-i / Halko Momoi (*Taiko no Tatsujin* song)
6. Junpaku Sanctuary (純白サンクチュアリィ) / Minori Chihara (*Kiddy Grade 2* ED)
7. Kimi ga Kureta Ano Hi (君がくれたあの日) / Minori Chihara
8. Koi no Dice Oshiete (恋のDice◻教えて) / Cy-Rim rev. (*Circusland I* theme song)
9. Seishoujou Ryouiki (聖少女領域) / Arika Takarano, Nana Mizuki (*Rozen Maiden* OP)
10. Tenohira (てのひら) / Naozumi Takahashi
11. Kimi ni Aete Yokatta (君に会えてよかった) / Naozumi Takahashi (*Radio Animelo Mix* theme song)
12. Yell！ / Minami Kuribayashi (*Super Robot Wars Original Generation: Divine Wars* ED)
13. Tsubasa wa Pleasure Line (翼はPleasure Line) / Minami Kuribayashi (*Chrono Crusade* OP)
14. Musouka (夢想歌) / Suara (*Utawarerumono* OP)
15. Kimi ga tame (キミガタメ) / Suara (*Utawarerumono* ED)
16. Float～Sora no Kanata e～ (Float～空の彼方で～)/ Tomoe Ohmi (*SoltyRei* ED)
17. Rock River e (ロックリバーへ) / Tomoe Ohmi (*Araiguma Rascal* OP)
18. Ushiroyubisasaregumi (うしろゆびさされ組) / Minami Kuribayashi, Halko Momoi (*High School! Kimengumi* OP)
19. Systematic Fantasy / m.o.v.e (*Gekisou! GT* ED)
20. Gamble Rumble / m.o.v.e (*Initial D* Third Stage OP)

21. Let's Go! Onmyouji (レッツゴー! 陰陽師) / Yabeno Hikomaru (Denchu) & Kotohime (MOMOEIKA) with Bouzu Dancers
22. Tokusou Sentai Dekaranger (特□□隊デカレンジャー) / Psychic Lover (*Tokusou Sentai Dekaranger* OP)
23. XTC / Psychic Lover (*Witchblade* OP)
24. Anata ga ita Mori (あなたがいた森) / Jyukai (*Fate/stay night* ED)
25. Sakasete wa Ikenai Hana (咲かせてはいけない花) / Jyukai
26. Yuukyou Seishunka (勇侠青春謳) / ALI PROJECT (*Code Geass* ED)
27. Ankoku Tengoku (暗□天国) / ALI PROJECT (*Kamichama Karin* OP)
28. Hizamazuite Ashi o Oname (跪いて足をお嘗め) / ALI PROJECT (*Princess Resurrection* ED)
29. Medley / Yoko Takahashi

    - Tamashii no Refrain (魂のルフラン) (*Evangelion: Death and Rebirth* theme song)
    - Zankoku na Tenshi no These (残酷な天使のテーゼ) (*Neon Genesis Evangelion* OP)

30. Bouken Desho Desho? (冒険でしょでしょ?) ~punk version~ / Aya Hirano (The Melancholy of Haruhi Suzumiya OP)
31. Break Out / JAM Project (*Super Robot Wars Original Generation: Divine Wars* OP)
32. VICTORY / JAM Project
33. SKILL / JAM Project
34. Justice to Believe / Nana Mizuki (*Wild Arms 5* OP)
35. SECRET AMBITION / Nana Mizuki (*Magical Girl Lyrical Nanoha Strikers* OP)
36. Heart-shaped chant / Nana Mizuki (*Shining Wind* Game OP)
37. Generation-A / Anisama Friends

    -encore-
38. OUTRIDE/ Anisama Friends
39. Generation-A / All performers

Media

CD

Generation-A

    Label: Dwango

    Release date: June 20, 2007

    Lyrics, composition: Masami Okui

    Arrangement: MACARONI□, Monta

    Animelo Summer Live 2007 Generation-A theme song

DVD

Animelo Summer Live 2007 -GENERATION A-

    Label: Lantis

    Release date: November 28, 2007

*Psychic Lover's Dekaranger performance is not included in DVD

2008

- Title: **Animelo Summer Live 2008 -Challenge-**
- Date: August 30 and August 31, 2008
- Location: Saitama Super Arena

- Sponsors: Dwango, QR
- Support: Kids Station, Television Kanagawa, FM NACK5, Saitama Super Arena
- Collaboration: Avex Trax, King Records, Geneon, Lantis

Performers

August 30

- Ali Project
- Yoko Ishida
- Karen Girl's
- GRANRODEO
- CooRie
- Minami Kuribayashi
- savage genius
- Suara
- Yukari Tamura
- Minori Chihara
- AAA
- Nana Mizuki
- m.o.v.e
- yozuca*
- Masami Okui
- AKINO from bless4

August 31

- ave;new feat. Saori Sakura
- Chiaki Ishikawa
- ELISA
- Kurobara Hozonkai
- Psychic Lover
- JAM Project
- Domestic Love Band
- The Idolmaster (Eriko Nakamura, Asami Imai, Chiaki Takahashi, Asami Shimoda)
- Aya Hirano
- Aki Misato
- May'n
- MOSAIC.WAV
- Haruko Momoi
- Chihiro Yonekura
- miko
- Lia
- Yoshiki Fukuyama
- Sound Horizon

Set List

Day 1 - August 30.

1. Koiseyo Onnanoko ~ Anone Mamimume☐Mogacho / Nana Mizuki + Yukari Tamura (*Gokujou Seitokai* OP & *Mamimume☐Mogacho* OP)
2. Douwa Meikyuu / Yukari Tamura (*Otogi-Jushi Akazukin* OP)
3. Bambino Bambina / Yukari Tamura (*Itazura Kuro Usagi* OP)
4. Melon no Theme - Yukari Oukoku Koka / Yukari Tamura
5. Strike Witches - Watashi ni Dekiru Koto / Yoko Ishida (*Strike Witches* OP)
6. Eien no Hana / Yoko Ishida (*Ai yori Aoshi* OP)
7. Sakura Saku Mirai Koi Yume / Yozuca* (*Da Capo* OP)
8. Morning - sugar rays / Yozuca*
9. Sentimental / CooRie (*Midori Days* OP)
10. Sonzai / CooRie (*Da Capo* ED)
11. DIVE INTO STREAM / m.o.v.e (*Initial D EXTREME STAGE* OP)
12. Gamble Rumble / m.o.v.e (*Initial D Third Stage* OP)
13. ZERO / AAA (*World Destruction* OP)
14. Climax Jump / AAA (*Kamen Rider Den-O* OP)
15. Over The Future / Karen Girl's (*Zettai Karen Children* OP)
16. nowhere / savage genius + Minori Chihara + Yozuca* (*Madlax* IN)
17. JUST TUNE / savage genius (*Yozakura Quartet* OP)
18. Omoi wo Kanadete / savage genius (*Uta Kata* OP)
19. Sousei no Aquarion / AKINO from bless4 (*Sousei no Aquarion* OP1)
20. Go Tight! / AKINO from bless4 (*Sousei no Aquarion* OP2)
21. Doukoku no Ame / GRANRODEO (*Koisuru Tenshi Angelique* OP)
22. Kenzen na Honnou / GRANRODEO
23. Love Jump / Minami Kuribayashi (*Kure-nai* OP)
24. Next Season / Minami Kuribayashi (*Kimi ga Nozomu Eien: Next Season* OP)
25. Shining☐Days / Minami Kuribayashi (*Mai-HiME* OP)
26. haunting Melody / Suara (*Tears to Tiara* PS3 OP)
27. Seiza / Suara (*Kusari* ED)
28. Shijin no Tabi from Contact / Minori Chihara
29. Ameagari no Hana Yo Sake / Minori Chihara
30. Yuki, Muon, Madobe Nite. / Minori Chihara (*The Melancholy of Haruhi Suzumiya* Yuki Nagato Character Song)
31. Rondo-revolution (輪舞-revolution) / Minori Chihara + Masami Okui (*Shoujo Kakumei Utena* OP ~ tribute album ver)
32. INSANITY / Masami Okui (*Muv-Luv Alternative Total Eclipse* OP)
33. Kotodama / ALI PROJECT (*Shigofumi* OP)
34. Ai to Makoto / ALI PROJECT (*pop'n music* Game song)
35. Waga Routashi Aku no Hana / ALI PROJECT (*Code Geass R2* ED)
36. Zankou no Gaia / Nana Mizuki (*SelectionX* ED)
37. Dancing in the velvet moon / Nana Mizuki (*Rosario + Vampire* ED)
38. Pray / Nana Mizuki (*Magical Girl Lyrical Nanoha Strikers* IN)
39. ETERNAL BLAZE / Nana Mizuki + Takarano Arika (*Magical Girl Lyrical Nanoha A's* OP)
40. Yells ～It's a beautiful life～ / Anisama 2008 artists (Animelo 2008 theme song)

Encore:

1. Generation-A / Anisama 2008 artists (Animelo 2007 theme song)
2. Yells ～It's a beautiful life～ / Anisama 2008 artists

Day 2 - August 31.

1. Omoide wa Okkusenman! / JAM Project + Aki Misato (*Mega Man 2* background song)
2. BLOOD QUEEN / Aki Misato (*Kaibutsu Oujo* OP)
3. Kimi ga Sora Datta / Aki Misato (*Mai-HiME* ED)
4. euphoric field / ELISA (*Ef: A Tale of Memories.* OP)
5. HIKARI / ELISA (*Nabari no Ou* ED)
6. Happy□Material / Domestic Love Band (*Mahō Sensei Negima!* OP)
7. Shangri-La / Domestic Love Band (*Soukyuu no Fafner* OP)
8. Asa to Yoru no Roman / Sound Horizon
9. Dorei Shijou / Sound Horizon
10. Seisen no Iberia (medley) / Sound Horizon
11. Hanabi ～ Manten Planetarium (medley) / Kurobara Hozonkai
12. Hikari / Kurobara Hozonkai (*Inukami!* OP)
13. Feel so Easy! / Haruko Momoi (*Mission-E* ED)
14. LOVE.EXE / Haruko Momoi (*Baldr Force Exe* OP)
15. Tenbatsu! Angel Rabbie / UNDER17 + MOSAIC.WAV (*Tenbatsu Angel Rabbie* ED)
16. SaikyouO×Keikaku / MOSAIC.WAV (*Sumomomo Momomo* OP)
17. Gacha Gacha Cute. Figu@MATE / MOSAIC.WAV (*Figu@MATE* OP)
18. true my heart ～ Iris ～ (medley) / ave;new feat. Saori Sakura (*Nursery Rhyme* OP)
19. Lovely□Angel!! (Short Ver.) / ave;new feat. Saori Sakura (*Stealing money Tenshi Twin Angel* song)
20. Marisa wa Taihen na Mono wo Nusunde Ikimashita / miko (IOSYS) (*Touhou Project*)
21. Tori no Uta / Lia (*Air* OP)
22. GO MY WAY!! ～ Kiramekirari ～ relations ～ Agent Yoru wo Yuku ～ Do-Dai ～Aoi Tori ～ GO MY
    WAY!! (medley) / Eriko Nakamura, Asami Imai, Chiaki Takahashi, Asami Shimoda from *The Idolmaster*
23. THE IDOLM@STER / Eriko Nakamura, Asami Imai, Chiaki Takahashi, Asami Shimoda from *The Idolmaster*
24. Makka na Chikai / Yoshiki Fukuyama (*Busou Renkin* OP)
25. Northern Cross / May'n (*Macross F* ED)
26. Iteza Gogo Kuji Don't be late / May'n (*Macross F* IN)
27. Precious Time, Glory Days / PSYCHIC LOVER (*Yu-Gi-Oh! GX* OP)
28. Kodou ～get closer～ / PSYCHIC LOVER (*Witchblade* IN)
29. Love Gun / Aya Hirano
30. Unnamed world / Aya Hirano (*Nijū Mensō no Musume* ED)
31. Uninstall / Chiaki Ishikawa (*Bokurano* OP)
32. Prototype / Chiaki Ishikawa (*Gundam 00* ED)
33. Anna Ni Issho Datta No Ni ～ Arashi no Naka de Kagayaite / Chiaki Ishikawa + Chihiro Yonekura (*Gundam
    SEED* ED & *Gundam The 08th MS Team* OP)
34. Eien no Tobira / Chihiro Yonekura (*Gundam The 08th MS Team Movie* OP)
35. FRIENDS / Chihiro Yonekura (*Senkaiden Houshin Engi* ED)
36. No Border / JAM Project
37. Rocks / JAM Project (*Super Robot Wars Original Generation* OP)
38. SKILL / JAM Project (*2nd Super Robot Wars Alpha* OP)
39. Yells ～It's a beautiful life～ / Anisama 2008 artists (Animelo 2008 theme song)

Encore:

1. OUTRIDE / Anisama 2008 artists (Animelo 2006 theme song)

2. Yells ∼It's a beautiful life∼ / Anisama 2008 artists

Media

CD

Yells ~It's A Beautiful Life~

>Label: Dwango / evolution

>Release date: July 23, 2008

>Lyrics: Masami Okui

>Composition: Hironobu Kageyama

>Animelo Summer Live 2008 Challenge theme song

DVD / Bluray

Animelo Summer Live 2008 -Challenge- 8.30

Animelo Summer Live 2008 -Challenge- 8.31

>Label: King Records

>Release date: March 25, 2009

※Some performances are not included in the release:

- Strike Witches - Watashi ni Dekiru Koto / Ishida Yoko
- All Sound Horizon and The Idolm@ster performances

2009

- Title: **Animelo Summer Live 2009 -RE:BRIDGE-**
- Date: August 22 and August 23, 2009
- Location: Saitama Super Arena
- Sponsors: Dwango, Nippon Cultural Broadcasting, Good Smile Company, Bushiroad
- Support: Animax Broadcast Japan, NACK5, Saitama Super Arena
- Collaboration: King Records, Geneon Entertainment, 5pb., flying DOG, Lantis...

Performers

August 22

- Ayane
- ALI PROJECT
- angela
- Chiaki Ishikawa
- Kanako Itō
- ELISA
- Minami Kuribayashi
- GRANRODEO
- Asami Shimoda
- JAM Project
- Minori Chihara
- Shoko Nakagawa (Special Appearance)
- Miku Hatsune (from Vocaloid 2)

- Beat Mario
- Yui Horie
- Manzo
- Mamoru Miyano
- May'n
- Haruko Momoi

August 23

- Tomoe Ohmi
- Kenji Ohtsuki and Zetsubou Shōjo Tachi (Ai Nonaka, Ryōko Shintani, Yū Kobayashi, Miyuki Sawashiro)
- Masami Okui
- GACKT
- Hironobu Kageyama
- Psychic Lover
- Yui Sakakibara
- savage genius
- Suara
- Yukari Tamura
- The Idolmaster (Eriko Nakamura, Asami Imai, Mayako Nigo)
- Aya Hirano
- FictionJunction
- Faylan
- Nana Mizuki
- m.o.v.e
- Yosei Teikoku
- Chihiro Yonekura

Set List

Day 1 - August 22.

1. Zankoku na Tenshi no These / angela + Chiaki Ishikawa (*Neon Genesis Evangelion* OP)
2. Spiral / angela (*Asura Cryin'* OP)
3. Shangri-La / angela (*Soukyuu no Fafner* OP)
4. F.D.D / Kanako Itou (*Chaos;Head* OP)
5. Tsuisou no Despair / Kanako Itou (*Higurashi no Naku Koro ni: Kizuna* OP)
6. Ai no Medicine ～Anisama Version～ / Haruko Momoi (*Nurse Witch Komugi-chan Magikarte* OP)
7. WONDER MOMO-i ～World tour version～ / Haruko Momoi (*Taiko no Tatsujin* song)
8. My Pace Daioh / Haruko Momoi + manzo (*Genshiken* OP)
9. Mizonokuchi Taiyou Zoku / manzo (*Tentai Senshi Sunred* OP)
10. Zoku Mizonokuchi Taiyou Zoku / manzo (*Tentai Senshi Sunred* OP)
11. Endless Tears... / Ayane (*11eyes Crossover* OP)
12. Complex Image / Ayane (*Higurashi no Naku Koro ni: Matsuri* OP)
13. Wonder Wind / ELISA (*Hayate no Gotoku!!* OP)
14. ebullient future ～Anisama Version～ / ELISA (*Ef: A Tale of Melodies.* OP)
15. Kokoro / Asami Shimoda
16. Miku Miku ni Shite Ageru ♪ / Miku Hatsune
17. Black Rock Shooter / Miku Hatsune
18. Help me, ERINNNNNN!! / Beat Mario (COOL&CREATE) (Touhou Project)

19. tRANCE / GRANRODEO (*Kurokami: The Animation* OP)

20. modern strange cowboy / GRANRODEO (*Needless* OP)

21. JET!!～Vanilla Salt (medley) / Yui Horie (*Toradora!* ED)

22. YAHHO!! / Yui Horie (*Kanamemo* ED)

23. HAPPY□MATERIAL / Minori Chihara + Yui Horie (*Mahō Sensei Negima!* OP)

24. First Pain / Chiaki Ishikawa (*Element Hunters* OP)

25. Prototype / Chiaki Ishikawa (*Gundam 00* ED)

26. J□S / Mamoru Miyano (*Card Gakuen* ED)

27. Discovery / Mamoru Miyano (*Fushigi Yuugi Suzaku Ibun* OP)

28. Kimi Shinitamou Koto Nakare / May'n (*Shangri-La* OP)

29. Diamond Crevasse / May'n (*Macross Frontier* ED)

30. Miracle Upper WL / May'n + Masami Okui (*Ontama!* OP)

31. Sorairo Days / Shoko Nakagawa (*Tengen Toppa Gurren-Lagann* OP)

32. Namida no Tane, Egao no Hana / Shoko Nakagawa (*Tengen Toppa Gurren Lagann The Movie: The Spiral Stone Chapter* OP)

33. WHAT'S UP GUYS? / Minami Kuribayashi + Kishow Taniyama (*Sorcerer Hunters* OP)

34. Kitei no Tsurugi / ALI PROJECT (*Kurogane no Linebarrels* OP)

35. Senritsu no Kodomotachi / ALI PROJECT (*Phantom Requiem for The Phantom* OP)

36. Jigoku no Mon / ALI PROJECT (*Phantom Requiem for The Phantom* ED)

37. Voyager train / Minori Chihara (*Plus Anison* ED)

38. Tomorrow's chance / Minori Chihara (*RunRun LAN* OP)

39. Paradise Lost / Minori Chihara (*Ga-Rei Zero* OP)

40. Precious Memories / Minami Kuribayashi (*Kimi Ga Nozomu Eien* OP)

41. Namida no Riyuu / Minami Kuribayashi (*School Days* ED)

42. sympathizer / Minami Kuribayashi (*Kurokami: The Animation* OP)

43. Crest of Z's / JAM Project (*Super Robot Wars Z PS2* OP)

44. Shugojin -the guardian- / JAM Project (*Shin Mazinger Shougeki! Z-Hen* OP)

45. Rescue Fire! / JAM Project (*Tomica Hero Rescue Fire* OP)

46. SKILL! / JAM Project (*2nd Super Robot Wars Alpha* OP)

47. RE:BRIDGE ～Return to Oneself～ / Anisama 2009 artists (Animelo 2009 theme song)

Encore:

1. OUTRIDE / Anisama 2009 artists (Animelo 2006 theme song)

2. RE:BRIDGE ～Return to Oneself～ / Anisama 2009 artists

Day 2 - August 23.

1. DISCOTHEQUE～MonStAR / Nana Mizuki + Aya Hirano (*Rosario + Vampire Capu2* OP)&(*Ani eggs* ED)

2. Super Driver / Aya Hirano (*The Melancholy of Haruhi Suzumiya* OP)

3. Set me free / Aya Hirano (*Run Run LAN!* OP)

4. Yakusoku no Basho e / Chihiro Yonekura (*Kaleido Star* OP)

5. 10 YEARS AFTER / Chihiro Yonekura (*Mobile Suit Gundam: The 08th MS Team* ED)

6. Soshite Boku wa... / Yui Sakakibara (*Prism Ark* OP)

7. Koi no Honoo / Yui Sakakibara (*Kanokon* ED)

8. Tamakui / Yousei Teikoku (*Ga-Rei Zero* IN)

9. last Moment / Yousei Teikoku (*Mai-Otome* Game IN)

10. Hitotoshite Jiku ga Bureteiru / Kenji Ohtsuki & Zetsubou-shoujotachi (*Sayonara Zetsubou Sensei* OP)

11. Kuusou Rumba / Kenji Ohtsuki & Zetsubou-shoujotachi (*Zoku Sayonara Zetsubou Sensei* OP)

12. Ringo Mogire Beam! / Kenji Ohtsuki & Zetsubou-shoujotachi (*Zan Sayonara Zetsubou Sensei* OP)

13. CHA-LA HEAD-CHA-LA ~acoustic version~ / Hironobu Kageyama (*Dragon Ball Z* OP)

14. SOLDIER DREAM / Hironobu Kageyama (*Saint Seiya* OP)

15. Parallel Hearts / FictionJunction (*Pandora Hearts* OP)

16. Akatsuki no Kuruma / FictionJunction (*Gundam Seed* IN)

17. nowhere / FictionJunction (*Madlax* IN)

18. Return to Love / Tomoe Ohmi (*Solty Rei* ED)

19. Maze / savage genius + Tomoe Ohmi (*Pandora Hearts* ED1)

20. Watashi o Mitsukete. / savage genius (*Pandora Hearts* ED2)

21. Dark Side of the Light / Faylan (*Ga-Rei Zero* IN)

22. mind as Judgment / Faylan (*Canaan* OP)

23. Tamashii no Rufuran / Masami Okui + Faylan (*Evangelion: Death and Rebirth* theme song)

24. LOVE SHIELD / Masami Okui (*Koisuru Otome to Shugo no Tate* Movie OP)

25. Maiochiru Yuki no You ni / Suara (*White Album* ED)

26. Free and Dream / Suara (*Tears to Tiara* OP)

27. DANZEN! Futari wa Precure / Yukari Tamura + Ryoko Shintani (*Futari wa Pretty Cure* OP)

28. LOST IN SPACE / Psychic Lover (*Tytania* ED)

29. THE IDOLM@STER / THE IDOLM@STER (*THE IDOLM@STER* OP)

30. Kiramekirari / THE IDOLM@STER

31. my song / THE IDOLM@STER

32. DOGFIGHT~Blazin' Beat (medley) / m.o.v.e (*Initial D Fourth Stage* OP)&(*Initial D Second Stage* OP)

33. Gravity / m.o.v.e (*Lucky Star* IN)

34. Chelsea Girl / Yukari Tamura (*Yukari Tamura, Mischievous Black Rabbit* OP)&(*Café Black Rabbit ~Secret Nook~* OP)

35. Tomorrow / Yukari Tamura (*Growlanser* OP)

36. Little Wish / Yukari Tamura (*Mahou Shoujo Lyrical Nanoha* ED)

37. Ai Senshi / GACKT (*Mobile Suit Gundam: Gundam vs. Gundam Next* OP)

38. The Next Decade / GACKT (*Kamen Rider Decade: All Riders vs. Dai-Shocker* Movie OP)

39. REDEMPTION / GACKT (*Dirge of Cerberus: Final Fantasy VII*)

40. Gimmick Game / Nana Mizuki + motsu (*Card Gakuen* OP)

41. Shin Ai / Nana Mizuki + Suara (*White Album* OP)

42. Etsuraku Camellia / Nana Mizuki (*Hammer laugh! One phrase* ED)

43. Orchestral Fantasia / Nana Mizuki (*MUSIC FIGHTER* POWER PLAY song)

44. RE:BRIDGE ~Return to Oneself~ / Anisama 2009 artists (Animelo 2009 theme song)

Encore:

1. ONENESS / Anisama 2009 artists (Animelo 2005 theme song)

2. RE:BRIDGE ~Return to Oneself~ / Anisama 2009 artists

Media

CD

RE:BRIDGE～Return to oneself～

Label: Dwango / Evolution

Release date: June 24, 2009

Lyrics: Masami Okui

Composition: Minami Kuribayashi

Animelo Summer Live 2009 RE:BRIDGE theme song

DVD / Bluray

Animelo Summer Live 2009 RE:BRIDGE 8.22

Animelo Summer Live 2009 RE:BRIDGE 8.23

Label: King Records

Release date: February 24, 2010

※Shoko Nakagawa's performances are not included in the release.

External links

- Animelo Summer Live 2005 -THE BRIDGE- [2]
- Animelo Summer Live 2006 -OUTRIDE- [3]
- Animelo Summer Live 2007 Generation-A [4]
- Animelo Summer Live 2008 -Challenge- [5]
- Animelo Summer Live 2009 -RE:BRIDGE- [1]
- Animelo Search Results on Akibanana.com [6]

References

[1] http://pc.animelo.jp/rebridge/
[2] http://pc.animelo.jp/bridge/
[3] http://pc.animelo.jp/outride/
[4] http://pc.animelo.jp/generation-a/
[5] http://pc.animelo.jp/challenge/
[6] http://www.akibanana.com/?q=search/node/animelo

Japan Aerospace Exploration Agency

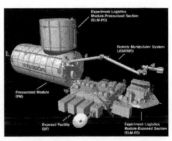

Owner	● Japan
Established	October 1, 2003 *(Successor agency to NASDA 1969-2003, ISAS 1981–2003 and NAL 1955–2003)*
Headquarters	Chōfu, Tokyo
Primary spaceport	Tanegashima Space Center
Motto	One JAXA
Administrator	Keiji Tachikawa
Budget	¥225 (USD 2.15) billion (FY2005)[1]
Website	www.jaxa.jp [2]

The **Japan Aerospace Exploration Agency** (独立行政法人宇宙航空研究開発機構 *Dokuritsu-gyōsei-hōjin Uchū Kōkū Kenkyū Kaihatsu Kikō*, literally "Independent Administration on the Exploration and Aviation of Space Study and Development Organization"), or **JAXA**, is Japan's national aerospace agency. JAXA was formed on October 1, 2003, as an Independent Administrative Institution through the merger of three previously independent organizations. JAXA is responsible for research, development and launch of satellites into orbit, and is fundamentally involved in many missions such as asteroid exploration and a possible human mission to the Moon.[3] Its motto is *One JAXA*[4] and corporate message is *Reaching for the skies, exploring space.*[5]

History

On October 1, 2003, three organizations were merged to form the new JAXA: Japan's **Institute of Space and Astronautical Science** (or ISAS), the **National Aerospace Laboratory of Japan** (NAL), and **National Space Development Agency of Japan** (NASDA).

Before the merger, ISAS was responsible for space and planetary research, while NAL was focused on aviation research. NASDA, which was founded on October 1, 1969, had developed rockets, satellites, and also built the Japanese Experiment Module, of which two of three sections have been added to the International Space Station.[6] The old NASDA headquarters were located at the current site of the Tanegashima Space Center, on Tanegashima Island, 115 kilometers south of Kyūshū. NASDA also trained Japanese astronauts, who flew with the US Space Shuttles.[7]

JAXA Kibo, the largest module for the ISS

Rockets

JAXA uses the H-IIA (H "two" A) rocket from the former NASDA body to launch engineering test satellites, weather satellites, etc. For science missions like X-ray astronomy, JAXA has been using the M-V ("Mu-five") solid-fueled rocket from the former ISAS. Additionally, JAXA is developing together with IHI, United Launch Alliance, and Galaxy Express Corporation (GALEX), the GX rocket. The GX will be the first rocket world wide to use liquefied natural gas (LNG) as the propellant. For experiments in the upper atmosphere JAXA uses the SS-520, S-520, and S-310 sounding rockets.

Success so far

Prior to the establishment of JAXA, ISAS had been most successful in its space program in the field of X-ray astronomy during the 1980s and 90s. Another successful area for Japan has been Very Long Baseline Interferometry (VLBI) with the HALCA mission. Additional success was achieved with solar observation and research of the magnetosphere, among other areas.

NASDA was mostly active in the field of communication satellite technology. However, since the satellite market of Japan is completely open, the first time a Japanese company won a contract for a civilian communication satellite was only in 2005. Another prime focus of the NASDA body is Earth climate observation.

JAXA was awarded the Space Foundation's John L. "Jack" Swigert, Jr., Award for Space Exploration in 2008.[8]

Launch development and missions

Rocket History

Japan launched its first satellite Ōsumi in 1970 with the L-4S rocket by ISAS. Unlike solid fueled rockets, Japan chose a much slower path with liquid fueled rocket technology. In the beginning NASDA used American models in license. The first model developed in Japan was the H-II introduced in 1994. However at the end of the 90s with two H-II launch failures, Japanese rocket technology came under criticism.

Early H-IIA missions

Japan's first space mission under JAXA, an H-IIA rocket launch on November 29, 2003, ended in failure due to stress problems. After a 15 month hiatus, JAXA performed a successful launch of an H-IIA rocket from Tanegashima Space Center, placing a satellite into orbit on February 26, 2005.

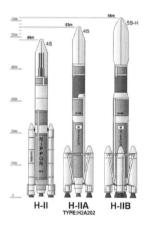

H-IIA & H-IIB

Lunar and Interplanetary Missions

Japan's first missions beyond Earth orbit were the 1985 Halley comet observation satellites Suisei and Sakigake. To prepare for future mission, ISAS tested Earth swing by orbits with the Hiten mission in 1990. The first Japanese interplanetary mission was the Mars Orbiter Nozomi (Planet-B), which was launched in 1998. It reached its target in 2003, but orbit injection had to be given up. Currently interplanetary missions remain at the ISAS group under the JAXA umbrella. However for FY 2008 JAXA is planning to set up an independent working group within the organization. New head for this group will be Hayabusa project manager Kawaguchi. [9] **Active Mission:** Hayabusa, SELENE, **Under Development:** Planet-C, BepiColombo, Hayabusa 2?

Small Body Exploration: Hayabusa mission

On May 9, 2003, Hayabusa (meaning, Peregrine falcon), was launched from an M-V rocket. The goal of this mission is to collect samples from a small near-Earth asteroid named 25143 Itokawa. The craft was scheduled to rendezvous in November 2005, and return to Earth with samples from the asteroid by July 2007. It was confirmed that the spacecraft successfully landed on the asteroid on November 20, 2005, after some initial confusion regarding the incoming data. On November 26, 2005, Hayabusa succeeded in making a soft contact, but whether it gathered the samples or not is unknown. Hayabusa is slated to return to earth in 2010.

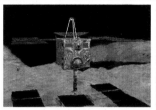

Hayabusa

For details see Hayabusa, Hayabusa 2

Solar sail research

On August 9, 2004, ISAS successfully deployed two prototype solar sails from a sounding rocket. A clover type sail was deployed at 122 km altitude and a fan type sail was deployed at 169 km altitude. Both sails used 7.5 micrometer thick film.

ISAS tested a solar sail again as a sub payload to the Astro-F (Akari) mission on February 22, 2006. However the solar sail did not deploy fully. ISAS tested a solar sail again as a sub payload of the Solar-B launch at September 23 2006, but contact with the probe was lost. The goal is to have a solar sail mission to Jupiter after 2010.

Lunar Explorations

After Hiten in 1990, ISAS planned a lunar exploration mission LUNAR-A but after delays due to technical problems, the project was terminated in January 2007. The seismometer penetrator design for Lunar-A may be reused in future mission.

On September 14, 2007, JAXA succeeded in launching lunar orbit explorer "*Kaguya*", also known as SELENE (costing 55 billion yen including launch vehicle), the largest such mission since the Apollo program, on an H-2A rocket. Its mission is to gather data on the moon's origin and evolution. It entered into a lunar orbit on October 4, 2007.[10] [11]

Astronomy Program

The first Japanese astronomy mission was x-ray satellite Hakucho (Corsa-B), which was launched in 1979. Later ISAS moved into solar observation, radio astronomy through Space VLBI and infrared astronomy. **Active Mission:** Suzaku, Akari, Hinode **Under Development:** ASTRO-G, ASTRO-H

Infrared astronomy

Japan's first infrared astronomy mission was the 15 cm IRTS telescope which was part of the SFU multipurpose satellite in 1995. IRTS scanned during its one month lifetime around 7% of the sky before SFU got brought back to Earth by the Space Shuttle. During the 1990s JAXA also gave ground support for the ESA Infrared Space Observatory (ISO) infrared mission.

AKARI (Astro-F)

The next step for JAXA was the AKARI spacecraft, with the pre-launch designation ASTRO-F. This satellite was launched on 21 February 2006. Its mission is infrared astronomy with a 68 cm telescope. This is the first all sky survey since the first infrared mission IRAS in 1983. (A 3.6 kg nanosatellite named CUTE-1.7 was also released from the same launch vehicle.) [12]

JAXA is also doing further R&D for increasing the performance of its mechanical coolers for its future infrared mission SPICA. This would enable a warm launch without liquid helium. SPICA has the same size as the ESA Herschel Space Observatory mission, but is planned with a temperature of just 4.5 K to be much colder. The launch is planned for the year 2015, however the mission is not yet fully funded. Also ESA and NASA might contribute an instrument each. [13]

For details see AKARI, IRTS.

X-ray astronomy

Starting from 1979 with Hakucho (CORSA-B), Japan achieved for nearly 20 years continuous observation with its Hinotori, Tenma, Ginga and Asuka (ASTRO-A to D) x-ray observation satellites. However in the year 2000 the launch of Japan's fifth x-ray observation satellite ASTRO-E failed (as it failed at launch it never received a

ASTRO-E

proper name). Then on July 10, 2005, JAXA was finally able to launch a new X-ray astronomy mission named Suzaku (ASTRO-E II). This launch was important for JAXA, because in the five years since the launch failure of the original ASTRO-E satellite, Japan was without an x-ray telescope. Three instruments were included in this satellite: an X-ray spectrometer (XRS), an X-ray imaging spectrometer (XIS), and a hard X-ray detector (HXD). However, the XRS was rendered inoperable due to a malfunction which caused the satellite to lose its supply of liquid helium.

The next planned x-ray mission is the MAXI all-sky X-ray scanner. It will continuously monitors astronomical X-ray objects over a broad energy band (0.5 to 30 keV). MAXI will be installed on the Japanese external module of the ISS. [14] After this mission JAXA plans to launch ASTRO-H, also known under the name NeXT, in the summer of 2013.

For details see ASTRO-E II (Suzaku). ASTRO-H

Solar astronomy

Japan's solar astronomy started in the early 80s with the launch of the *Hinotori* (ASTRO-A) x-ray mission. The Hinode (SOLAR-B) spacecraft, the follow-on to the Japan/US/UK Yohkoh (SOLAR-A) spacecraft, was launched on 23 September 2006. [15] [16] A SOLAR-C can be expected sometime after 2010. However no details are worked out yet other than it will not be launched with the former ISASs Mu rockets. Instead H-2A from Tanegashima could launch it. As H-2A is more powerful SOLAR-C could either be heavier or be stationed at L_1 (Lagrange point 1).

For details see Hinode.

Radio Astronomy

In 1998 Japan launched the HALCA (Muses-B) Mission, the world first spacecraft dedicated to create SPACE VLBI observations of Pulsars among others. To do so, ISAS set up a ground network around the world through international cooperation. The observation part of the mission lasted until 2003 and the satellite was retired at the end of 2005. In FY 2006 Japan funded the ASTRO-G as the succeeding mission. Launch is planned for FY 2012.

For details see:

ASTRO-G HALCA

Technology Tests

One of the primary duties of the former NASDA body was the testing of new space technologies, mostly in the field of communication. The first test satellite was ETS-I,launched in 1975. However during the 1990s NASDA was hit by bad luck with the problems surrounding the ETS-VI and COMETS missions. Nevertheless testing of communication technologies remains as one of the Jaxas key duties in cooperation with NICT. **Active Mission:** ETS-VIII, WINDS, Index **Under Development:** QZSS-1 **Retired:** OICETS

ETS-VIII and WINDS

To upgrade Japans communication technology the Japanese state launched the i-Space initiative with the ETS-VIII and WINDS missions.[17]

ETS-VIII was launched on December 18 2006. The purpose of ETS-VIII is to test communication equipment with two very large antennas and an atomic clock test. On December 26 both antennas were successfully deployed. This didn´t come unexpected, since JAXA tested the deployment mechanism before with the LDREX-2 Mission, which was launched on October 14 with the European Ariane 5. The test was successful. The mission of WINDS is to create the worlds fastest satellite internet connection. WINDS was launched in February 2008.

For details see ETS-VIII, WINDS

OICETS and INDEX

On August 24, 2005, JAXA launched the experimental satellites OICETS and INDEX with the Dnepr rocket. OICETS mission is to test optical links with the European Space Agency (ESA) satellite ARTEMIS, which is around 40,000 km away from OICETS. The experiment was successful on December 9, when the link could be established. In March 2006 Jaxa could establish with OICETS the worldwide first optical links between a LEO satellite and a ground station first in Japan and in June 2006 with a mobile station in Germany.

For details see OICETS

INDEX is a small 70 kg satellite for testing various equipment and for a small aurora observation mission. The satellite is currently in the extended mission phase.

For details see INDEX

Earth Observation Programme

Japan's first Earth observation satellites were MOS-1a and MOS-1b launched in 1987 and 1990. During the 1990s and the new millennium this programme came under heavy fire, because both Adeos (Midori) and Adeos 2 (Midori 2) satellites failed after just 10 months in orbit.

Active Mission: ALOS, GOSAT **Under Development:** GCOM-W, GCOM-C, ALOS 2 SAR

ALOS

In January 2006, JAXA successfully launched the Advanced Land Observation Satellite (ALOS/Daichi). Communication between ALOS and the ground station in Japan will be done through the Kodama Data Relay Satellite, which was launched during 2002. This project is under intense pressure due to the shorter than expected life time of the ADEOS II (Midori) Earth Observation Mission. For the following on mission JAXA plans to split the mission into a radar satellite and an optical satellite. ALOS 2 SAR is currently planned for the winter of FY 2012.

MTSAT-1

Rainfall Observation

Since Japan is an island nation and gets struck by typhoons every year, research about the dynamics of the atmospheric is a very important issue. For this reason Japan launched in 1997 the TRMM mission in cooperation with NASA, to observe the tropical rainfall seasons. JAXA and NASA are planning a successor to the TRMM mission. However because of NASA budget problems the launch date of the GPM project got pushed back to the year 2013. For further research NASDA although launched the ADEOS and ADEOS II missions in 1996 and 2003. However due to various reasons both satellites had a much shorter than expected life term.

Monitoring of carbon dioxide

At the end of FY 2008 JAXA launched the satellite GOSAT (Greenhouse Gas Observing SATellite) to help scientists determine and monitor the density distribution of carbon dioxide in the atmosphere. The satellite is being jointly developed by JAXA and Japan's Ministry of the Environment. JAXA is building the satellite while the Ministry is in charge of the data that will be collected. Since the number of ground-based carbon dioxide observatories cannot monitor enough of the world's atmosphere and are distributed unevenly throughout the globe, the GOSAT may be able to gather more accurate data and fill in the gaps on the globe where there are no observatories on the ground. Sensors for methane and other greenhouse gasses are also being considered for the satellite, although the plans are not yet finalized. The satellite weighs approximately 1650 kg and is expected to have a life span of 5 years.

GCOM series

Next funded earth observation mission after GOSAT is the GCOM earth observation programme as a successor to ADEOS II (Midori) and the Aqua mission. To reduce the risk and for a longer observation time the mission will be split into smaller satellites. Altogether GCOM will be a series of six satellites. First launch, GCOM-W is scheduled for February 2012 with the H-IIA. Second launch GCOM-C is currently planned for February 2014.

For details see: ADEOS, ADEOS II, TRMM, GPM, GOSAT, GCOM-W

Satellites for other agencies

For weather observation Japan launched on February 2005 the Multi-Functional Transport Satellite 1R (MTSAT-1R). The success of this launch was critical for Japan, since the original MTSAT-1 couldn't be put into orbit because of a launch failure with the H-2 rocket in 1999. Since then Japan relied for weather forecasting on an old satellite which was already beyond its useful life term and on American systems. On February 18, 2006, JAXA, as head of the H-IIA at this time, successfully launched the MTSAT-2 aboard a H-2A rocket. MTSAT-2 is the backup to the MTSAT-1R. The MTSAT-2 uses the DS-2000 satellite bus developed by Mitsubishi Electric.[18] The DS-2000 is also used for the DRTS Kodama, ETS-VIII and the Superbird 7 communication satellite, making it the first commercial success for Japan.

As a secondary mission both the MTSAT-1R and MTSAT-2 help to direct air traffic.

Other JAXA satellites currently in use

- Exos-D (Akebono) Aurora Observation, since 1989.
- GEOTAIL magnetosphere observation satellite (since 1992)
- DRTS (Kodama) Data Relay Satellite, since 2002. (Projected Life Span is 7 years)

On going joint missions with NASA are the Tropical Rainfall Measuring Mission (TRMM), the Aqua Earth Observation Satellite.

Finished Missions

- OICETS, Technology Demonstration 2005-2009 (retired)
- SELENE, Moon probe 2007-2009 (retired)
- Micro Lab Sat 1, Small engineering mission, launch 2002. (retired 27 September 2006)
- HALCA, Space VLBI 1997-2005 (retired)
- Nozomi, Mars Mission 1998-2003 (failed)
- MDS-1, Technology Demonstration 2002-2003 (retired)
- ADEOS 2, (Midori 2) Earth Observation 2002-2003 (lost)

Future missions

As JAXA shifted away from international efforts beginning in 2005, plans are developing for independent space missions, such as a proposed manned mission to the moon.

HTV-1

2009 and beyond

On February 23, 2008 JAXA launched the Wideband InterNetworking engineering test and Demonstration Satellite (WINDS), also called "KIZUNA." WINDS will facilitate experiments with faster internet connections. The launch, using H-IIA launch vehicle 14, took place from the Tanegashima Space Center.[19]

On September 10, 2009 the first H-IIB rocket was successfully launched, delivering the HTV-1 freighter to resupply the International Space Station.[20]

Another project is the Global Precipitation Measurement/Dual-frequency Precipitation Radar (GPM/DPR) which is a joint development with NASA. This mission is the successor to the highly successful TRMM mission. JAXA will develop the radar and provide the launch vehicle. Other countries/agencies like China, India, ESA etc. will provide the subsatellites. The aim of this mission is to measure global rainfall. However because of NASA budget limitations this project was pushed back to 2010.

In the year 2009 JAXA plans to launch the first satellite of the Quasi Zenith Satellite System (QZSS), a subsystem of the global positioning system (GPS). Two others are expected to follow later. If successful, one satellite will be in a zenith position over Japan full time. The QZSS mission is the last scheduled major independent mission for JAXA, as no major civilian projects were funded after that for now. The only exception is the IGS programme which will be continued beyond 2008. However it seems Japan is pressing forward now with the GCOM earth observation satellites as successors to the ADEOS missions. First launch is planned for 2010. In 2009 Japan also plans to launch a new version of the IGS with an improved resolution of 60 cm.

Launch schedule

First launch of the H-IIB and the HTV is September 1, 2009. After the first flight one HTV launch is planned during each FY until 2015. (If not mentioned otherwise launch vehicle for the following missions is the H-IIA.)

FY 2010

- Quasi Zenith Satellite System
- PLANET-C, probe to Venus, launch: May, 2010

FY 2011

- GCOM-W, Climate Observation satellite, launch: Feb, 2012

FY 2012

- ALOS 2 SAR, Earth Observation satellite, launch: Winter 2012
- ASTRO-G (VSOP-2) successor to the Halca mission, launch: Summer 2012
- TOPS Telescope Observatory for Planets on Small-satellite launch Feb, 2012 (First launch of the new Advanced Solid Rocket, the successor to the M-V.

FY 2013

- GPM, successor to the TRMM joint NASA mission
- BepiColombo, joint ESA mission to Mercury, launch: 2013 (LV: Ariane 5)
- ASTRO-H x-ray observatory, launch: summer 2013.
- GCOM-C, Climate Observation satellite, launch: Feb, 2014

Other missions

For the 2012 ESA EarthCare mission, JAXA will provide the radar system on the satellite. JAXA is also providing the Light Particle Telescope(LPT) for the 2008 Jason 2 satellite by the French CNES. JAXA will provide the Auroral Electron Sensor (AES) for the Taiwanese FORMOSAT-5.[21]

- SmartSat-1, small communication test and sun corona observation, Mission status unclear
- XEUS joint X-Ray telescope with ESA, launch after 2015.
- Sohla-2 Small PETSAT Demonstration Satellite

New orientation of JAXA

Planning interstellar research missions can take up to seven years, such as the ASTRO-E. Due to the lag time between these interstellar events and mission planning time, opportunities to gain new knowledge about the cosmos might be lost. To prevent this, JAXA plans on using smaller, faster missions from 2010 onwards. JAXA is also planning to develop a new solid fueled rocket to replace the twelve year old M-V.

Developing Projects

- IKARUS (Interplanetary Kite-craft Accelerated by Radiation Of the Sun), a small size powered-solar sail experimental spacecraft for Jupiter and Trojan asteroids exploration.

Future plans

- Selene-2, a moon landing mission
- Hayabusa 2, for launch in 2010-2011 for target 1999JU3
- Hayabusa Mk2/Marco Polo
- Human Lunar Systems, conceptual system study on the future human lunar outpost
- ALOS 2, earth observation
- SPICA, a 3,5 meter infrared telescope to be placed at L2
- JASMINE, infrared telescope for measuring the universe
- DIOS, small scale x-ray observation

Human Space Program

Japan has not yet developed its own manned spacecraft and is not currently developing one. Sometime ago an unmanned space shuttle HOPE-Xproject launched by conventional space launcher H-II was developed for several years but was postponed. Then the simpler manned capsule Fuji was proposed but not adopted. Projects of single-stage to orbit, reusable launch vehicle horizontal takeoff and landing ASSTS and vertical takeoff and landing Kankoh-maru also exist but have not been adopted.

The first Japanese citizen to fly in space was Toyohiro Akiyama, a journalist sponsored by TBS, who flew on the Soviet Soyuz TM-11 in December 1990. He spent more than seven days in space on the Mir Space station, in what the Soviets called their first commercial spaceflight which allowed them to earn $14 million. The first professional Japanese astronaut was Mamoru Mohri, a NASDA astronaut who flew his first space mission aboard the STS-47 mission in 1992.

Under a new plan, JAXA has set a goal of constructing a manned lunar base in 2030. Astronauts would be sent to the Moon by beyond 2020 which is approximately the same time as Indian Space Research Organisation (ISRO) manned lunar mission beyond 2020, China National Space Administration (CNSA) manned lunar mission in 2030 and NASA's Project Constellation plans to return to the Moon in 2019 with its Orion-Altair project) so that they will start construction of the base to be completed by 2030.[22]

Before this Moon goals JAXA intends to develop the manned spacecraft launched by space launcher H-IIB [23]

Supersonic aircraft development

Besides the H-IIA and M-5 rockets, JAXA is also developing technology for a next-generation supersonic transport that could become the commercial replacement for the Concorde. The design goal of the project (working name NEXST) is to develop a jet that can carry 300 passengers at Mach 2. A subscale model of the jet underwent aerodynamic testing in September and October 2005 in Australia. [24] The economic success of such a project is still unclear, and as a consequence the project has been met with limited interest from Japanese aerospace companies like Mitsubishi Heavy Industries so far.

Reusable Launch Vehicles

Until 2003 JAXA (ISAS) conducted research on a reusable launch vehicle under the Reusable Vehicle Testing (RVT) project.

Research centers and offices

JAXA has research centers in many locations in Japan, and some offices overseas. Its headquarters are in Chōfu, Tokyo. It also has

- Earth Observation Research Center (EORC), Tokyo
- Earth Observation Center (EOC) in Hatayama
- Noshiro Testing Center (NTC) - Established in 1962. It carries out development and testing of rocket engines.
- Sanriku Balloon Center (SBC) - Balloons have been launched from this site since 1971.
- Kakuda Space Propulsion Center (KSPC) - Leads the development of rocket engines. Works mainly with development of liquid fuel engines.
- Sagamihara Campus (ISAS) - Development of experimental equipment for rockets and satellites. Also administrative buildings.
- Tanegashima Space Center
- Tsukuba Space Center (TKSC) in Tsukuba. This is the center of Japan's space network. It is involved in research and development of satellites and rockets, and tracking and controlling of satellites. It develops experimental

equipment for the Japanese Experiment Module ("Kibo"). Training of astronauts also takes place here. For International Space Station operations, the Japanese Flight Control Team is located at the Space Station Integration & Promotion Center (SSIPC) in Tsukuba. SSIPC communicates regularly with ISS crewmembers via S-band audio.[25]

- Uchinoura Space Center

Other space agencies in Japan

Not included into the JAXA organization is the Institute for unmanned space experiment free flyer (USEF), Japan´s other space agency.

See also

- Independent Administrative Institution (IAI), 2001
- List of Independent Administrative Institutions (Japan)

External links

- JAXA [2]
- "JAXA 2025" Presentation [26]
- "JAXA Channel" Official YouTube channel [27]
- International Space Station (ISS) and "Kibo" Information center [28]
- RAND Report on Japan's Space Program, 2005 [29]
- CSIS Report on US-Japan Space Policy Cooperation, 2003 [30]
- GOSAT satellite [31]

These three links are archived sites of the JAXA predecessor agencies:

- NASDA [32]
- ISAS [33]
- NAL [34]

References

[1] "IV. 決算報告書 (Balance Report)" (http://www.jaxa.jp/about/finance/pdf/finance_17-04.pdf) (in Japanese) (PDF). 平成17事業年度 JAXA財務諸表等に関する事項. JAXA. 2006-08-30. . Retrieved 2007-06-29.
[2] http://www.jaxa.jp/index_e.html
[3] "Japan launches biggest moon mission since Apollo landings" (http://www.guardian.co.uk/science/2007/sep/15/spaceexploration.japan). guardian.co.uk/science. . Retrieved 2007-09-16.
[4] Keiji Tachikawa - JAXA in 2006 (http://www.jaxa.jp/article/interview/vol21/p2_e.html)
[5] Jaxa | About Jaxa (http://www.jaxa.jp/about/index_e.html)
[6] "Consolidated Launch Manifest - Space Shuttle Flights and ISS Assembly Sequence" (http://www.nasa.gov/mission_pages/station/structure/iss_manifest.html). NASA. .
[7] Kamiya, Setsuko, " Japan a low-key player in space race (http://search.japantimes.co.jp/cgi-bin/nn20090630i1.html)", Japan Times, June 30, 2009, p. 3.
[8] http://www.nationalspacesymposium.org/symposium-awards
[9] http://ilws.gsfc.nasa.gov/china_jaxa.pdf
[10] Japancorp.net, Japan Successfully Launches Lunar Explorer "Kaguya" (http://www.japancorp.net/Article.Asp?Art_ID=15429)
[11] BBC NEWS, Japan launches first lunar probe (http://news.bbc.co.uk/2/hi/asia-pacific/6994272.stm)
[12] http://nssdc.gsfc.nasa.gov/database/MasterCatalog?sc=2006-005A
[13] http://www.jaxa.jp/article/interview/no19/p4_e.html
[14] http://iss.sfo.jaxa.jp/kibo/kibomefc/maxi_e.html
[15] http://solar-b.nao.ac.jp/index_e.shtml
[16] http://solar-b.msfc.nasa.gov/
[17] http://i-space.jaxa.jp/ispace.html

[18] "製品のご紹介 製品・衛星プラットフォーム／DS2000" (http://www.mitsubishielectric.co.jp/society/space/products/platform_b.
html) (in Japanese). Mitsubishi Electric. . Retrieved 2008-08-03.

[19] "Launch Result of the KIZUNA (WINDS) by the H-IIA Launch Vehicle No. 14 (H-IIA F14)" (http://www.jaxa.jp/press/2008/02/
20080223_kizuna_e.html). JAXA. .

[20] "Japan's space freighter in orbit" (http://news.bbc.co.uk/2/hi/science/nature/8249357.stm). *Jonathan Amos*. BBC. 2009-08-10. .
Retrieved 2009-09-10.

[21] http://www.pssc.ncku.edu.tw/FISFES/Presentation/FISFES_2008-11(Hirahara).pdf

[22] Staff Writers (August 3, 2006). "Japan Plans Moon Base By 2030" (http://www.moondaily.com/reports/
Japan_Plans_Moon_Base_By_2030_999.html). *Moon Daily*. SpaceDaily. . Retrieved 2006-11-17.

[23] http://www.flightglobal.com/blogs/hyperbola/2009/01/pictures-jaxas-h-iib-launched.html

[24] http://news.yahoo.com/s/ap/20051010/ap_on_sc/supersonic_jet

[25] "ISS On-Orbit Status 04/23/09" (http://www.hq.nasa.gov/osf/iss_reports/reports2009/04-23-2009.htm). NASA. .

[26] http://www.jaxa.jp/about/vision_missions/long_term/jaxa_vision_e.pdf

[27] http://www.youtube.com/jaxachannel

[28] http://iss.jaxa.jp/en/

[29] http://www.rand.org/publications/TR/TR184/

[30] http://www.csis.org/media/csis/pubs/taskforcereport.pdf

[31] http://www.jaxa.jp/missions/projects/sat/eos/gosat/index_e.html

[32] http://www.nasda.go.jp/index_e.html

[33] http://www.isas.ac.jp/e/index.shtml

[34] http://www.nal.go.jp/Welcome-e.html

Article Sources and Contributors

Vocaloid *Source*: http://en.wikipedia.org/w/index.php?oldid=341734199 *Contributors*: 54ggg, Alton.arts, Antonio Lopez, Aster Selene, Avianwind, Blacksaingrain, Bnynms, Bokura1000, Chase-san, Chicojava, Ciaccona, Comrade Pajitnov, DJCruithne, Deadkid dk, Debresser, Defragged, Dhilvert, Dinoguy1000, Eky-w-, FUCKDAMNANN, Fife5000, Fortifiedchicken, Fractyl, Fred Hsu, Graf Bobby, Grutness, HappyDesu1, Hatsune Miku, HumanGir, Inuzukaluv, Jacob Poon, Jake Wartenberg, Jellypuzzle, Jicksmashbros, JoleneCarterLee, Jpark3909, Juhachi, JuniorB03, Justiflake, Kanesue, KatTheVampire, King of Hearts, Kingpin13, Kokokun, KratosAuiron, KyuuA4, L-Zwei, La Pianista, LauraOrganaSolo, Li Jianliang, Lololollol53, LuminousPath, Magnius, Malambis, Manuke, Marcika, Masoris, MatthewVanitas, Michal Nebyla, Midori, Mightyfastpig, Mikkumiku, MikuMaker, Mimiru1985, Miyuki, Moocowsrule, Mopza, Morio, Mzm5zbC3, Nagle, Oherman, OlEnglish, P. S. Burton, Parrothead1983, Peter S., Pmsyyz, Princesspaula487, Psholic, Rcjsuen, RepublicanJacobite, Revth, Rizelon, Rwwww, Ryulong, SaxTeacher, Sepiraph, Shii, Siawase, Signumyagami, Supallcomm, Tally Solleni, Tetsuya takashi, TheKhakinator, Thebestandtheworst, Tracy santiago27, Un1234l, Uncle Dick, Venustas 12, Vocaloiduser, WikHead, WindOwl, Wonchop, Wyatt915, 446 anonymous edits

Yamaha Corporation *Source*: http://en.wikipedia.org/w/index.php?oldid=341172020 *Contributors*: **mech**, 16@r, 7, Acela Express, Acroterion, Ahmad87, Aitias, AliaGemma, Andete, AndonicO, Andycjp, Andyman100, Angela, Anthropic42, Aodhdubh, ArglebargleIV, Arjun01, Armando82, Arthana, Autocracy, Bbakerxyz, Billy Bishop, Binksternet, Blackpudding, Bobbart1, BobtheEditorMan, Bomen168, Bongomatic, Bradisley, Bunsenhoneydew, CGorman, Can't sleep, clown will eat me, Canyouhearmenow, Carl.bunderson, Cat910, Catgut, Cglassey, Chaparral2J, Chirags, Ciphers, Closeapple, Courthut, Cremepuff222, Cybermonsters, DAJF, DMacks, Da Vynci, Dabomb87, Danmayna, David5150, DavidFarmbrough, Ddstretch, Dfrg.msc, DmitryKo, DocWatson42, Dogbreath, Domin12345, Doseiai2, Dr G, DrFod, Drmies, Ds13, Dyl, EJF, Eastmain, Ehudshapira, Endroit, Escape Orbit, EvanSeeds, Ewlyahoocom, Fanoftheworld, Fataltourist, Fdevilbis, Femto, Feral-Golduck, Fingon1, Flamurai, Flibble, Fosnez, Frodet, Furrykef, Gaius Cornelius, Garykahn, Gentgeen, GlassCobra, GlassFET, Glen, GreyCat, Grm wnr, Grutness, Gustav von Humpelschmumpel, Haakon Thue Lie, Harvester, Her-my-o-knee, HexaChord, Hooperbloob, Hut 8.5, Ian3055, Ief, Inks002, Inzy, Ixfd64, J.delanoy, Jacj, Jacoplane, Jake Wartenberg, Jamcib, James Blue Jazz, Jbrock1, Jechasteen, JoeyJoJoShabbaduJr, Joseph Solis in Australia, Josephrr92, Jpatokal, KFP, Kanwar rajan, Karenjc, Kazrak, Kbh3rd, Kcchiefs202, Kilo-Lima, Kintetsubuffalo, Kmccoy, KnowledgeOfSelf, Kur0, LG4761, Lenilucho, Lewys93, Liftarn, Lombroso, Lost tourist, Lova Falk, LtPowers, ML1986, Magetoo, Malcohol, Marko sk, Menchi, Methoxodus, Mike Sorensen, MonkeyCMonkeyDo, Mrceleb2007, Msxfan, Mtffm, Mwakin21, Myscrnnm, Mütze, Naniwako, Neverender 899, Nixdorf, Oarih, Oliviosu, Organs plus, Ourai, Panoptical, Patrick@moorevyas.com, Pavel.petrov, Peewack, Philip Trueman, Phoenix10k, Phyllis1753, Pixel8, Pmaguire, Podzemnik, Psr450, Puerto07, Rabit, Radagast83, RadicalBender, Raneksi, Renegadeviking, Retiono Virginian, Rettetast, Rhe br, Rjwilmsi, Robert.Harker, RobyWayne, Rockfang, Roguegeek, Roscoe x, RugbyUrulez, SarekOfVulcan, SaxTeacher, Scepia, Schroederrt, SharShar, Sionus, Sirimiri, Sjones23, Someguy1221, Sprinter76, Sriharsha4, Stevage, Steveshelokhonov, Suduser85, Superbeecat, TastyPoutine, Tedius Zanarukando, The Thing, The Thing That Should Not Be, Thedjatclubrock, Thehornet, Thingg, Tigers boy, Tlotoxl, Tornvmax, Tregoweth, Two Bananas, Van helsing, Vegaswikian, Vivek04, Vmadeira, WJetChao, Walkmanwalkman, Weissmann, Wheat, Wiki edit Jonny, Willking1979, Wysdom, Xen 1986, Yas, Yodaat, Yotcmdr, 281 anonymous edits

Lyrics *Source*: http://en.wikipedia.org/w/index.php?oldid=341477420 *Contributors*: -x-dannii-x-, 13alexander, 16@r, 5theye, ABF, ARC Gritt, Academic Challenger, Adam Bishop, Adavidw, Ageekgal, Al.locke, Alexfrance250291, AlistairMcMillan, Altermike, Andycjp, Anthony5429, Antonio Lopez, Apoc2400, Apparition11, ArielGold, Art LaPella, Asterion, Asxvideos, Baim78, Bernis, Bhamv, Blade76, Blm07, Bobet, Bobo192, Brianga, Bunnyhop11, CFCF, Can't sleep, clown will eat me, Canterbury Tail, Captain-tucker, Carinemily, CaseyIsCool, Cdc, Cedders, Cethegus, Charles Matthews, Chase me ladies, I'm the Cavalry, ChrisCork, Contact@music-free-download.net, CoramVobis, CryptoDerk, Csmaster2005, Damian Yerrick, Dantadd, DavidRF, Deeplogic, Defunkt, Deltabeignet, Diaby, Diddi, Dragoburaggo, Dylan Lake, ERcheck, EamonnPKeane, Eclecticology, Elipongo, Elonka, Eloquence, Emmedenney, Enviroboy, Enzo Aquarius, Epbr123, Eric the Rexman, Fennec, Flatfoot4444, Friedenbach, Frogfusious, Garo, Gimme danger, GregAsche, Hadal, HappyCamper, Henry Flower, Hu12, Hyacinth, II MusLiM HyBRiD II, IP 84.5, Iglam, J.delanoy, Jacklee, Jbinder, Jeggish105, Jennavecia, Jerry, Jklin, Jls33fsls, Jmundo, John254, JohnCub, Johnnyw, Jorunn, Jwy, Kairosis, Kazvorpal, Kent Wang, Kerii57, Ks0stm, Kukini, La Pianista, Lafraia, Lathspell, Mac, Macrazy, Madder, Mandarax, Mark.deane, MarkBollett, Martin451, Max Naylor, Merenta, Minesweeper, MinuteHand, Mister Floyd, Mrmanhattanproject, Mxipp, My name, N Shar, NCurse, Nakon, Nanenj, NawlinWiki, Nbrett1, Nemanjakron, Nightkey, No Guru, Notinasnaid, Nuggetman, OMGitsCTC, Osmosis, Paddles, Patstuart, Pavel Vozenilek, Pewwer42, Pibwl, Pietaster, Pinko1977, Pip2andahalf, Pkoden, Possum, Ptanham, Pyrope, Quill, RMFan1, RadiantRay, RedWolf, Reflex Reaction, Rholton, Rls, Robert Foley, RockMFR, RoyBoy, Rrburke, Ryulong, SColombo, Saberwyn, Sannse, Sean Patrick Griffiths, Seriema, SilkTork, Soliloquial, Sridharinfinity, Ss112, Stephen4800, Stereotek, Struway2, Supastabi, TUF-KAT, Tedder, Thatdog, The Thing That Should Not Be, TheMadBaron, Tigers boy, Tiptoety, Toktosunov, Topbanana, Trysha, Unmesh.bhosle, Vinithehat, Vxlover, Wayland, WebJunkie, Weyes, Willirennen, Willking1979, WinterSpw, Wpbmma, Wtmitchell, Yachtsman1, Yumegusa, Yvil, Zazaban, Zgvozden, 410 anonymous edits

Melody *Source*: http://en.wikipedia.org/w/index.php?oldid=339809496 *Contributors*: 1dragon, 210.49.20.xxx, Adashiel, Alansohn, Altadena, Andres, Andycjp, Archanamiya, Ary29, Ayrton Prost, Bannana123, Beano, Beinel, Beland, Bemoeial, Blanchardb, CALR, Cahk, Can't sleep, clown will eat me, Capricorn42, Ched Davis, Chuck Sirloin, Clark Kimberling, Cremepuff222, Defenseman Emeritus, Denny, DougsTech, Duncharris, Dúnadan, El C, Emhoo, Ewlyahoocom, Face, Fenneck, Fuhghettaboutit, Gaius Cornelius, Glenn, Grace Note, H1es-, Hadal, Helixblue, Hellmaggot, Hyacinth, Ilya, Iridescent, J.delanoy, JLvanOs, Janamills, JayAston2009, Jengirl1988, Jfurr1981, Jim, Joseph Solis in Australia, Juliancolton, Just64helpin, Kamaki, Khalid hassani, Kimyu12, Leafyplant, Leonus, Limideen, Lishea, Lordotis, Martin451, Maxim, Mcbill88, Megan.rw1, Merphant, Michael Hardy, MuffledThud, Munci, Nathanael Bar-Aur L., Natural Cut, Nekura, Nemomelody, Ninja Wizard, Oda Mari, Omnipaedista, OriginalJay, Oxymoron83, PGWG, Pearle, Philip Trueman, Phillipsacp, Pingveno, Pol098, R. S. Shaw, Radimast, Red Winged Duck, Redheylin, Reflex Reaction, Regibox, Rigaudon, Roachgod, RobertG, Romanm, Ryulong, Sam Hocevar, Sam Weller, Schaefer, SchfiftyThree, Sfan00 IMG, Sinneed, Sluzzelin, SmartGuy, Smokizzy, Smoove Z, Stephen Henry Davies, Summer Song, Tarquin, Taurrandir, The Archivist, The Cat and the Owl, Thingg, Thivierr, Timneu22, Tlork Thunderhead, Trevor MacInnis, Tumble-Weed, Tuneman42, Undead warrior, VinceBowdren, Vlmastra, WhisperToMe, WhiteWizard42, Widders, Wik, Yaddle34, Yamamoto Ichiro, 215 anonymous edits

Miriam Stockley *Source*: http://en.wikipedia.org/w/index.php?oldid=332831534 *Contributors*: Alai, Andykgoss, Blanche Hunt, Candice, Candyfloss, Crystallina, Dl2000, Dom Kaos, Ettlz, Garion96, Harry Hayfield, Hovea, Japanese Searobin, LegalTide, Monni1995, Orlandocalling, Owen, Peter S., Pharos, RexNL, Rich Farmbrough, Rwelliot, Sjones23, Sopoforic, Sposato, Tilla, TubularWorld, Urmelbeauftragter, Valentinian, WereSpielChequers, Xnux, トリノ特許許可局, 41 anonymous edits

NAMM Show *Source*: http://en.wikipedia.org/w/index.php?oldid=339900306 *Contributors*: Aeternus, BarryWood, BillyRamirez, Brandonp@namm.com, Cbarbry, CharlesGillingham, Choster, Denniss, EricBarbour, Googuse, Houtlijm, JaGa, Jpdavidson, Kalel2007, Kevinjohnstone, Leahtwosaints, Mboverload, Niteowlneils, Paul J Stevens, Pearle, Peter S., PhilKnight, Redevent, Sabrebattletank, Shawnc, Sonett72, TexasAndroid, Tremspeed, 41 anonymous edits

Computer music *Source*: http://en.wikipedia.org/w/index.php?oldid=340354452 *Contributors*: "alyosha", 0, 137.112.129.xxx, 213.253.40.xxx, Angelus1753, AnnaFrance, Appraiser, Atoll, Backtable, Belbernard, Bijanzelli, CambridgeBayWeather, Capricorn42, Chris 73, Closedmouth, Conversion script, Dan56, Davidqwerty, Davigoli, Digego, Djulio, Droidus, Furrykef, Geldryk, Gewang, Gilbertevich, Guitarmasterclass, Heron, Jbartman, Jerome Kohl, Joswig, Jwolf, Karol Langner, Khalfani khaldun, Logan, M7, Magic Window, Marc Mongenet, MarylandArtLover, Matterson52, Mcld, Meiticheol, Merphant, Midicontest, Milkmandelivery, MinorContributor, Nebula2357, Neonarcade, Ninly, Nixeagle, Omegatron, Optimale, Oscillon, Oystein, Parsifal, Physicistjedi, Pierre cummings, Pkirlin, Polluks, Quiddity, Quinobi, R.123, Requestion, Rev3rend, RichardVeryard, Rjwilmsi, Rmkeller, Ronz, Savetz, Sdbeck, Semitransgenic, Shadowjams, Sigma 7, Skyezx, Sluzzelin, Stefanos Leon, TJRC, Tabletop, TedColes, Tfine80, Tgies, Thumbuki, Toineheuvelmans, Trivialist, Twang, Userafw, Valueyou, Versageek, Vivenot, Viznut, Woohookitty, Xezbeth, Zundark, Zvika, 129 anonymous edits

Speech synthesis *Source*: http://en.wikipedia.org/w/index.php?oldid=340887082 *Contributors*: 12 Noon, 67-21-48-122, 7, A3 nm, Aaronrp, Abdull, Ace Frahm, Actam, Agoubard, Altermike, Andreas Bischoff, AndrewHowse, Angr, Argon233, Arj, Arvindn, BStrategic, Back ache, Backstabb, Badly Bradley, Beetstra, Betterusername, BjarteSorensen, Bobbo, Bobo192, Bocharov, Bovineone, Brianreading, Brion VIBBER, Bumm13, CHIPSpeaking, Calltech, Calton, Caltrop, Canderra, Canis Lupus, CanisRufus, Cassowary, Chachou, Charlie danger, Chmod007, Chocolateboy, Chris the speller, Chrischris349, Chrysoula, Chuck Sirloin, ChuckOp, Conscious, CortlandKlein, Crystallina, Damian Yerrick, Darkspartan4121, Dave w74, David Gerard, Ddp224, Dennishc, Discospinster, Dogman15, Doodle77, Dr.K., Dwiki, Dycedarg, Dylan Lake, E946, EdC, Eik Corell, Ellamosi, Eurleif, Everyking, Fr33kman, Fractaler, Ftiercel, Furrykef, Gaius Cornelius, Gary Cziko, Gerstman ny, Giftlite, Glenn, Grendelkhan, GreyCat, Gwalla, H, Hamedkhoshakhlagh, Harryboyles, Heron, Hhanke, Hike395, Holizz, Instine, Intgr, Invitatious, Itman, Itzcuauhtli, JJblack, JLaTondre, Jacob Poon, Jeff Henderson, Jenliv2ryd, Jimfbleak, Jimich, Jlittlet, Joelr31, JoergenB, JohnMRye, Johnny Au, K7aay, Kaldosh, Kate, Kayemel3, Kcordina, Kindall, Knowledgerend, Kuszi, Kwamikagami, Kwekubo, KyraVixen, L736E, Lakshmish, Lalalele, Latitude0116, Ligulem, Lksdfvbmwe, Lukeluke1978, Lupin, Mac, MarcS, Marskell, Martinevans123, Matrixbob, Matt Crypto, MatthPeder, Maury Markowitz, MaviAteş, Michal Jurosz, MonteChristof, Morfeuz, MotherFerginPrincess, Moyogo, MrTree, MuthuKutty, N1RK4UDSK714, Nahitk, Naikrovek, Nanard, Neelix, NeonMerlin, Nige7, Nlu, Nohat, Oli Filth, Outriggr, Oznull, Oznux, PAR, Palthrow, Patrick, Paulson74, Paxse, Pb30, Pedro, Pengo, Pgillman, Politepunk, Polluks, Ppareit, PrimroseGuy, Python eggs, Quadell, Raffaele Megabyte, Raul654, RedWolf, Redox12, Renaudforestie, Requestion, Rich Farmbrough, Rjanag, Rjwilmsi, Rmcguire, RobertG, Roberto111199, Rogerb67, Ruud Koot, Santhosh.thottingal, Satori, Savetz, Saxbryn, Scott Sanchez, Serezniy, Shadowjams, Shaftesbury12, Shaneymac, SigmaEpsilon, Silas S. Brown, Simeon, Singlebarrel, SiobhanHansa, Smack, Sonjaaa, Southpolekat, Speechgrl, Stephanos Georgios John, StephenPratt, Stephenchou0722, SteveRenals, Supremeknowledge, Suruena, T33guy, TMC1221, TextToSpeech, The Founders Intent, Thefreethinker, Thenickdude, Thesilverbail, Thoobik, Thorenn, Thue, Thw416, TimMagic, Tjwood, Tlesher, Tobias Bergemann, Tommy Blueseed, Tony1, Trainra, Twikir, Twistor, Twthmoses, Uluboz, ValerieChernek, Versus22, Viajero, Voxsprechen, W4HTX, W9000, Wayne Miller, Wayp123, Weheh, Wesley crossman, Wik, Wikipodium, Wizzard2006, Wolf grey, Xdenizen, Yamamoto Ichiro, Yiddophile, Yosri, Yrithinnd, Ziusudra, चंद्रकांत धुतडमल, 410 anonymous edits

Nico Nico Douga *Source*: http://en.wikipedia.org/w/index.php?oldid=339693217 *Contributors*: 16@r, Aitias, Bdonlan, Cheetahmen2, Deadkid dk, Infrogmation, Jpark3909, Juhachi, Keito, Kyoww, LightFlare, Lita5dozen, Littlebtc, Mahal Aly, Makesdark, Megaman10, Metagold, Miyuki, Mmuuishikawa, Moocowsrule, Neuname, Nike787, Niyaniya, Nopira, Pitan, Rcjsuen, Rjanag, Shii, Shinya0526, Swind, Tedmund, Wonchop, Woohookitty, 66 anonymous edits

Loituma Girl *Source*: http://en.wikipedia.org/w/index.php?oldid=336293508 *Contributors*: 54together, AGK, Addict 2006, Alexius08, Alphachimp, Anaraug, Andycjp, Antoshi, Aphemushroom, ArglebargleIV, Arjun01, Arteyu, BalkanFever, BerserkShinji01, Bobo192, Brat32, Brian Kendig, Buchanan-Hermit, Can't sleep, clown will eat me, Captain Cornflake, ChickenSoda X4, Chrisxx, ComFritter18, Crashmatrix, Ctrlshifttab, D, DanielCD, DarkKunai, DarkMasterBob, Darkmaster2004, Dekimasu, DerHexer, DiamondDragon, Dillard421, DocWatson42, Dominic, Doug, Dwight666, Dysepsion, Eagle creek, Ehn, Elemesh, Elwood j blues, Emigdioofmiami, Erachima, EvilBrak, Finlay McWalter, Flameviper, Flewis, Fox816, Fpmfpm, FreeKresge, Friendly Neighbour, Frigo, Froth, Furrykef, Furudenendu, GTBacchus, GhostStalker, Gilliam, Goober-peas, Gormon, Gudeldar, Gutworth, Haham hanuka, HamburgerRadio, Hkit, Hogtree, IPAddressConflict, Ianzcoolist, Illythr, Intranetusa, Iridescent, J.delanoy, J04n, Jedravent, Jeff G., JeongAhn, Joizashmo, Juhachi, Kamui04, Kazu-kun, Kitetsu, Korg, Kribbeh, Krisgrotius, Kubigula, Leuko, Lightlowemon, MRkukov, Melchoir, Melly42, Michael Hardy, MichaelFrey, Midnyt, Mild Bill Hiccup, Monni1995, MrBoo, Mysid, NeoVampTrunks, Neoyamaneko, Netrat, Night Gyr, Octane, Ohnoitsjamie, Omicronpersei8, Ospinad, Oxymoron83, PatrikR, Perlmangle, Persian Poet Gal, Piano non troppo, Pianotrees, Piroteknix, Poetic Decay, Pooh, PuzzletChung, RealGrouchy, Remurmur, Rentaferret, Repku, ReyBrujo, Rhobite, Ricky81682, Rock Kawaii, RockMFR, Roninbk, Royaljared, Ræv, SHARU(ja), ST47, Sakkmatt, Saxifrage, Schloob, SchuminWeb, Seifip, Senshi, Sephiroth BCR, Sharkeynoyz, Shay Guy, Shepanator, Shii, Sigmafactor, Sitonera, Snarfies, Spartaz, SquirrelJS667, Stephenb, Stimpy9337, StrangerAtaru, SunDragon34, Svick, Sylocat, Synchronism, T@nn, TCHJ3K, Tbone762, Tckma, The New Mikemoral, TheEmerald, Thingg, Timc, Toddst1, Trainra, Tree Biting Conspiracy, Tregoweth, Trialsanderrors, Triradiates, Tuuur, Tyrasibion, Unconscious, Venu62, Vianello, Viridae, Vishahu, Vusys, Wafulz, WereSpielChequers, WhisperToMe, WhoWhatWhenWhere, WikHead, Wikiman232, Wikipediatrix, Woohookitty, Wykypydya, Xalrun, Yancyfry jr, Yonghokim, ZeBoxx, ZeroOne, Zetawoof, Zscout370, 423 anonymous edits

Cameo appearance *Source*: http://en.wikipedia.org/w/index.php?oldid=340073460 *Contributors*: A Nobody, Adiel, After Midnight, AgentPeppermint, Ahoerstemeier, Ajo Mama, Alecsdaniel, Alton.arts, Angela, Antandrus, Aris Katsaris, Art LaPella, Aspects, Attilios, AussieLegend, Bachrach44, Bernhard Bauer, Bex-of-mcr, BigBadaboom0, Binabik80, Bobber1, Bobblehead, Bodachi, Bolt Vanderhuge, Bombyourself, Boorishbehaviour, Bovineone, Branddobbe, Bunnyhop11, CLW, Californian Treehugger, Camboy8, CanadianCaesar, CanisRufus, Capricorn42, Captain Crawdad, Cardsplayer4life, Cburnett, Cedars, Ceoil, Christopherlin, Colonies Chris, Cordd, Craverguy, Cubs Fan, Cwoyte, D.brodale, DJ Clayworth, Da Joe, Dabernat, Dafyddwrm, Daibhid C, DanteOrange, Darthdj31, Dbachmann, Diavolino79, Didgeman, Dijxtra, Dogcow, Dom Lochet, Dr bab, Drew3D, Duffbeerforme, Dutchmonkey9000, Easter Monkey, Ejk81, El Coro, Ellsworth, Emibird92, EngineerScotty, Eurosong, Face, Felicity4711, FinnFinn, Francis Schonken, Franz.87, Furrykef, GSK, GarethaRemembered, Ghosts&empties, Grammarmonger, Grandpafootsoldier, Granpuff, Great Scott, GusF, Harryema, Heimstern, HillValleyTelegraph, Huax, Hunterman1000, Hégésippe Cormier, I Feel Tired, Imaginationac, IstvanWolf, JJuran, JamesMLane, JamieS93, Jdelanoy, JeffHCross, Jeffrey O. Gustafson, JesseRafe, Jesster79, Jetfire85, Jim Huggins, Joeyconnick, JohnDBuell, JohnDoe0007, Jovianeye, Just H, Just64helpin, Karada, Kathar, Kbolino, Keeves, Kenny TM~, Kguirnela, Kindall, Kissaki, Kusunose, Lee M, Lefty, Leolaursen, Lestari, Liberatus, Lightmouse, Loadmaster, Longhair, Luckyz, Lumantu, Mark Reichert, MarmadukeFan85, Meamemg, Mecanismo, MegX, Mel Etitis, Melaen, Melodia, Merovingian, Miaow Miaow, Michael Dorosh, Michael Hardy, MightyJordan, Missmarple, Moogle, MrBoo, Mrguyguy226, NawlinWiki, Nidonocu, Nihil novi, Not a slave, NotAnotherAliGFan, NrDg, ONEder Boy, Obakeneko, Oerjan, Oleg Alexandrov, Optigan13, Orilux, PTSE, Paul A, Paul Abrahams, Pedant, Perks, Phil Boswell, Pixymeancatt, Pjdm91, Polylerus, Pressforaction, Project2501a, RainbowCrane, Reconsider the static, Red-Blue-White, Retodon8, RevRaven, Roadrunner, Roberta F., Roccondil, Roffnix, Rparle, SMB39A, Sam Hocevar, Scarletsmith, SchuminWeb, SidP, Silvermoonx, Sinblox, Sintonak.X, Sir Nicholas de Mimsy-Porpington, Sjorford, Skyraider, Slawojarek, Soramimi124, Sottolacqua, Soundguy99, SouthernNights, Sssoul, Stemonitis, Stormwatch, Str1977, Subwayguy, Sudastelaro, Sulvo, Swaq, The JPS, The Rambling Man, The Rogue Penguin, TheBearPaw, TheDJ, Thefishnamedcarl, Thinking Stone, ThrowingStick, Thu, Timc, Time for action, TinyMark, Trekphiler, Uthanc, VeryVerily, Vonvon, WadeSimMiser, Webdinger, Wetman, Wikidudeman, Willy11009, Woohookitty, Wuntzt, Xezbeth, Xy7, Zosomaniac, Zpb52, Zzuuzz, 352 anonymous edits

Victor Entertainment *Source*: http://en.wikipedia.org/w/index.php?oldid=341689560 *Contributors*: ACSE, Andycjp, Anonymous55, Art LaPella, Aspects, Chubbles, Cwbrown, DAJF, DBailey635, Daniel Christensen, Dpilat, Evanreyes, Ganryuu, Hatto, Iridescent, Kbuster, Ketsumaru, Lavateraguy, Magus Melchior, Mako098765, Martarius, Mewtwowimmer, Minlilin, Morishinichi, Nihonjoe, Queenmomcat, Rappyboy2008, Rcjsuen, Runrun 923, Slysplace, Sposato, Teddy.Coughlin, Tsubasa mokona, WereSpielChequers, Wikimankeith, Yui Tokito, 40 anonymous edits

Animelo Summer Live *Source*: http://en.wikipedia.org/w/index.php?oldid=339809615 *Contributors*: Addshore, Alansohn, Fratrep, Highwind888, Inclusivedisjunction, Jevansen, Jouttex, Juhachi, Kurorohunter, Mikamura, MouseImm, Nerefis, Pandacomics, Rich Farmbrough, Ryulong, ShelfSkewed, Simon Brady, Thanhtu87, Tydus Arandor, 85 anonymous edits

Japan Aerospace Exploration Agency *Source*: http://en.wikipedia.org/w/index.php?oldid=340004247 *Contributors*: A300st, Aafm, Akadruid, Alba, All Is One, Alsandro, Andrwsc, Andycjp, Aris Katsaris, Arria Belli, Arsonal, Ash sul, Astrowob, AvengerX, Bcat, Beland, Benbest, Bendono, Biblbroks, Bobblehead, Bobblewik, Bobo12345, Body-Hentai-Knocking, Brockert, Bryan Derksen, CES, Chesnok, Christian List, CiaPan, Circeus, Cla68, Cmdrjameson, Colby Peterson, Colincbn, Confuzion, Crazycane, Ctpm, Cursed Pretzel, DannyWilde, David Kernow, David Woodward, Denelson83, DerHexer, Djmckee1, Doady, DougsTech, Dutchsatellites.com, Easphi, Emperorthma, Endroit, Entropy=0, Eubulides, Fenice, Florentino floro, FlyHigh, Forezt, Freestylefrappe, Fukumoto, GRAND OUTCAST, GW Simulations, Gaara42, Gaijin Ninja, Gaius Cornelius, Gene Nygaard, Gits (Neo), Grim23, Harriv, Harsha850, Hektor, Hellisp, Henrygb, IanOsgood, Ignatzmice, Igossow, Inputpersona, Intgr, JTN, Jatkins, Jogloran, Jonathan66, Joseph Solis in Australia, Josh Parris, KGyST, Ke4roh, Keiya Chinen, Kintetsubuffalo, Koreakorea1, Kusunose, Lightmouse, MBK004, Mad031683, Marshallsumter, Maverick, McSly, Mgiganteus1, Mikus, Mlm42, Modulatum, Moocowsrule, Mugu-shisai, Necessary Evil, Neier, Neilc, Nesnad, Nikai, Nono64, Nrpf22pr, Nuno Tavares, Nv8200p, Ohms law, Oxymoron83, Perceval, Pfainuk, Pharaoh of the Wizards, Premkudva, Raso mk, Ricnun, Rillian, Rjwilmsi, Rmhermen, RuM, Rudykog, Ryulong, Sam Hocevar, Sardanaphalus, Sdsds, Sfg2, ShadowHntr, Shinkansen Fan, Shoeofdeath, Southsidejohnny, Spot87, Srain, Ssolbergj, Svick, Taichi, Techietim, Tenmei, The Woman Who Sold The World, Thedeadlypython, Tokek, Tul21, V8Cougar, Van der Hoorn, Voidvector, Vriullop, WereSpielChequers, WhiteDragon, WngLdr34, Xenon54, Zigger, 朝彦, 331 anonymous edits

Image Sources, Licenses and Contributors

License

GNU Free Documentation License Version 1.2, November 2002 Copyright (C) 2000,2001,2002 Free Software Foundation, Inc. 59 Temple Place, Suite 330, Boston, MA 02111-1307 USA Everyone is permitted to copy and distribute verbatim copies of this license document, but changing it is not allowed.

0. PREAMBLE

The purpose of this License is to make a manual, textbook, or other functional and useful document "free" in the sense of freedom: to assure everyone the effective freedom to copy and redistribute it, with or without modifying it, either commercially or noncommercially. Secondarily, this License preserves for the author and publisher a way to get credit for their work, while not being considered responsible for modifications made by others. This License is a kind of "copyleft", which means that derivative works of the document must themselves be free in the same sense. It complements the GNU General Public License, which is a copyleft license designed for free software. We have designed this License in order to use it for manuals for free software, because free software needs free documentation: a free program should come with manuals providing the same freedoms that the software does. But this License is not limited to software manuals; it can be used for any textual work, regardless of subject matter or whether it is published as a printed book. We recommend this License principally for works whose purpose is instruction or reference.

1. APPLICABILITY AND DEFINITIONS

This License applies to any manual or other work, in any medium, that contains a notice placed by the copyright holder saying it can be distributed under the terms of this License. Such a notice grants a world-wide, royalty-free license, unlimited in duration, to use that work under the conditions stated herein. The "Document", below, refers to any such manual or work. Any member of the public is a licensee, and is addressed as "you". You accept the license if you copy, modify or distribute the work in a way requiring permission under copyright law. A "Modified Version" of the Document means any work containing the Document or a portion of it, either copied verbatim, or with modifications and/or translated into another language. A "Secondary Section" is a named appendix or a front-matter section of the Document that deals exclusively with the relationship of the publishers or authors of the Document to the Document's overall subject (or to related matters) and contains nothing that could fall directly within that overall subject. (Thus, if the Document is in part a textbook of mathematics, a Secondary Section may not explain any mathematics.) The relationship could be a matter of historical connection with the subject or with related matters, or of legal, commercial, philosophical, ethical or political position regarding them. The "Invariant Sections" are certain Secondary Sections whose titles are designated, as being those of Invariant Sections, in the notice that says that the Document is released under this License. If a section does not fit the above definition of Secondary then it is not allowed to be designated as Invariant. The Document may contain zero Invariant Sections. If the Document does not identify any Invariant Sections then there are none. The "Cover Texts" are certain short passages of text that are listed, as Front-Cover Texts or Back-Cover Texts, in the notice that says that the Document is released under this License. A Front-Cover Text may be at most 5 words, and a Back-Cover Text may be at most 25 words. A "Transparent" copy of the Document means a machine-readable copy, represented in a format whose specification is available to the general public, that is suitable for revising the document straightforwardly with generic text editors or (for images composed of pixels) generic paint programs or (for drawings) some widely available drawing editor, and that is suitable for input to text formatters or for automatic translation to a variety of formats suitable for input to text formatters. A copy made in an otherwise Transparent file format whose markup, or absence of markup, has been arranged to thwart or discourage subsequent modification by readers is not Transparent. An image format is not Transparent if used for any substantial amount of text. A copy that is not "Transparent" is called "Opaque". Examples of suitable formats for Transparent copies include plain ASCII without markup, Texinfo input format, LaTeX input format, SGML or XML using a publicly available DTD, and standard-conforming simple HTML, PostScript or PDF designed for human modification. Examples of transparent image formats include PNG, XCF and JPG. Opaque formats include proprietary formats that can be read and edited only by proprietary word processors, SGML or XML for which the DTD and/or processing tools are not generally available, and the machine-generated HTML, PostScript or PDF produced by some word processors for output purposes only. The "Title Page" means, for a printed book, the title page itself, plus such following pages as are needed to hold, legibly, the material this License requires to appear in the title page. For works in formats which do not have any title page as such, "Title Page" means the text near the most prominent appearance of the work's title, preceding the beginning of the body of the text. A section "Entitled XYZ" means a named subunit of the Document whose title either is precisely XYZ or contains XYZ in parentheses following text that translates XYZ in another language. (Here XYZ stands for a specific section name mentioned below, such as "Acknowledgements", "Dedications", "Endorsements", or "History".) To "Preserve the Title" of such a section when you modify the Document means that it remains a section "Entitled XYZ" according to this definition. The Document may include Warranty Disclaimers next to the notice which states that this License applies to the Document. These Warranty Disclaimers are considered to be included by reference in this License, but only as regards disclaiming warranties: any other implication that these Warranty Disclaimers may have is void and has no effect on the meaning of this License.

2. VERBATIM COPYING

You may copy and distribute the Document in any medium, either commercially or noncommercially, provided that this License, the copyright notices, and the license notice saying this License applies to the Document are reproduced in all copies, and that you add no other conditions whatsoever to those of this License. You may not use technical measures to obstruct or control the reading or further copying of the copies you make or distribute. However, you may accept compensation in exchange for copies. If you distribute a large enough number of copies you must also follow the conditions in section 3. You may also lend copies, under the same conditions stated above, and you may publicly display copies.

3. COPYING IN QUANTITY

If you publish printed copies (or copies in media that commonly have printed covers) of the Document, numbering more than 100, and the Document's license notice requires Cover Texts, you must enclose the copies in covers that carry, clearly and legibly, all these Cover Texts: Front-Cover Texts on the front cover, and Back-Cover Texts on the back cover. Both covers must also clearly and legibly identify you as the publisher of these copies. The front cover must present the full title with all words of the title equally prominent and visible. You may add other material on the covers in addition. Copying with changes limited to the covers, as long as they preserve the title of the Document and satisfy these conditions, can be treated as verbatim copying in other respects. If the required texts for either cover are too voluminous to fit legibly, you should put the first ones listed (as many as fit reasonably) on the actual cover, and continue the rest onto adjacent pages. If you publish or distribute Opaque copies of the Document numbering more than 100, you must either include a machine-readable Transparent copy along with each Opaque copy, or state in or with each Opaque copy a computer-network location from which the general network-using public has access to download using public-standard network protocols a complete Transparent copy of the Document, free of added material. If you use the latter option, you must take reasonably prudent steps, when you begin distribution of Opaque copies in quantity, to ensure that this Transparent copy will remain thus accessible at the stated location until at least one year after the last time you distribute an Opaque copy (directly or through your agents or retailers) of that edition to the public. It is requested, but not required, that you contact the authors of the Document well before redistributing any large number of copies, to give them a chance to provide you with an updated version of the Document.

4. MODIFICATIONS

You may copy and distribute a Modified Version of the Document under the conditions of sections 2 and 3 above, provided that you release the Modified Version under precisely this License, with the Modified Version filling the role of the Document, thus licensing distribution and modification of the Modified Version to whoever possesses a copy of it. In addition, you must do these things in the Modified Version: A. Use in the Title Page (and on the covers, if any) a title distinct from that of the Document, and from those of previous versions (which should, if there were any, be listed in the History section of the Document). You may use the same title as a previous version if the original publisher of that version gives permission. B. List on the Title Page, as authors, one or more persons or entities responsible for authorship of the modifications in the Modified Version, together with at least five of the principal authors of the Document (all of its principal authors, if it has fewer than five), unless they release you from this requirement. C. State on the Title page the name of the publisher of the Modified Version, as the publisher. D. Preserve all the copyright notices of the Document. E. Add an appropriate copyright notice for your modifications adjacent to the other copyright notices. F. Include, immediately after the copyright notices, a license notice giving the public permission to use the Modified Version under the terms of this License, in the form shown in the Addendum below. G. Preserve in that license notice the full lists of Invariant Sections and required Cover Texts given in the Document's license notice. H. Include an unaltered copy of this License. I. Preserve the section Entitled "History", Preserve its Title, and add to it an item stating at least the title, year, new authors, and publisher of the Modified Version as given on the Title Page. If there is no section Entitled "History" in the Document, create one stating the title, year, authors, and publisher of the Document as given on its Title Page, then add an item describing the Modified Version as stated in the previous sentence. J. Preserve the network location, if any, given in the Document for public access to a Transparent copy of the Document, and likewise the network locations given in the Document for previous versions it was based on. These may be placed in the "History" section. You may omit a network location for a work that was published at least four years before the Document itself, or if the original publisher of the version it refers to gives permission. K. For any section Entitled "Acknowledgements" or "Dedications", Preserve the Title of the section, and preserve in the section all the substance and tone of each of the contributor acknowledgements and/or dedications given therein. L. Preserve all the Invariant Sections of the Document, unaltered in their text and in their titles. Section numbers or the equivalent are not considered part of the section titles. M. Delete any section Entitled "Endorsements". Such a section may not be included in the Modified Version. N. Do not retitle any existing section to be Entitled "Endorsements" or to conflict in title with any Invariant Section. O. Preserve any Warranty Disclaimers. If the Modified Version includes new front-matter sections or appendices that qualify as Secondary Sections and contain no material copied from the Document, you may at your option designate some or all of these sections as invariant. To do this, add their titles to the list of Invariant Sections in the Modified Version's license notice. These titles must be distinct from any other section titles. You may add a section Entitled "Endorsements", provided it contains nothing but endorsements of your Modified Version by various parties--for example, statements of peer review or that the text has been approved by an organization as the authoritative definition of a standard. You may add a passage of up to five words as a Front-Cover Text, and a passage of up to 25 words as a Back-Cover Text, to the end of the list of Cover Texts in the Modified Version. Only one passage of Front-Cover Text and one of Back-Cover Text may be added by (or through arrangements made by) any one entity. If the Document already includes a cover text for the same cover, previously added by you or by arrangement made by the same entity you are acting on behalf of, you may not add another; but

you may replace the old one, on explicit permission from the previous publisher that added the old one. The author(s) and publisher(s) of the Document do not by this License give permission to use their names for publicity for or to assert or imply endorsement of any Modified Version.

5. COMBINING DOCUMENTS

You may combine the Document with other documents released under this License, under the terms defined in section 4 above for modified versions, provided that you include in the combination all of the Invariant Sections of all of the original documents, unmodified, and list them all as Invariant Sections of your combined work in its license notice, and that you preserve all their Warranty Disclaimers. The combined work need only contain one copy of this License, and multiple identical Invariant Sections may be replaced with a single copy. If there are multiple Invariant Sections with the same name but different contents, make the title of each such section unique by adding at the end of it, in parentheses, the name of the original author or publisher of that section if known, or else a unique number. Make the same adjustment to the section titles in the list of Invariant Sections in the license notice of the combined work. In the combination, you must combine any sections Entitled "History" in the various original documents, forming one section Entitled "History"; likewise combine any sections Entitled "Acknowledgements", and any sections Entitled "Dedications". You must delete all sections Entitled "Endorsements".

6. COLLECTIONS OF DOCUMENTS

You may make a collection consisting of the Document and other documents released under this License, and replace the individual copies of this License in the various documents with a single copy that is included in the collection, provided that you follow the rules of this License for verbatim copying of each of the documents in all other respects. You may extract a single document from such a collection, and distribute it individually under this License, provided you insert a copy of this License into the extracted document, and follow this License in all other respects regarding verbatim copying of that document.

7. AGGREGATION WITH INDEPENDENT WORKS

A compilation of the Document or its derivatives with other separate and independent documents or works, in or on a volume of a storage or distribution medium, is called an "aggregate" if the copyright resulting from the compilation is not used to limit the legal rights of the compilation's users beyond what the individual works permit. When the Document is included in an aggregate, this License does not apply to the other works in the aggregate which are not themselves derivative works of the Document. If the Cover Text requirement of section 3 is applicable to these copies of the Document, then if the Document is less than one half of the entire aggregate, the Document's Cover Texts may be placed on covers that bracket the Document within the aggregate, or the electronic equivalent of covers if the Document is in electronic form. Otherwise they must appear on printed covers that bracket the whole aggregate.

8. TRANSLATION

Translation is considered a kind of modification, so you may distribute translations of the Document under the terms of section 4. Replacing Invariant Sections with translations requires special permission from their copyright holders, but you may include translations of some or all Invariant Sections in addition to the original versions of these Invariant Sections. You may include a translation of this License, and all the license notices in the Document, and any Warranty Disclaimers, provided that you also include the original English version of this License and the original versions of those notices and disclaimers. In case of a disagreement between the translation and the original version of this License or a notice or disclaimer, the original version will prevail. If a section in the Document is Entitled "Acknowledgements", "Dedications", or "History", the requirement (section 4) to Preserve its Title (section 1) will typically require changing the actual title.

9. TERMINATION

You may not copy, modify, sublicense, or distribute the Document except as expressly provided for under this License. Any other attempt to copy, modify, sublicense or distribute the Document is void, and will automatically terminate your rights under this License. However, parties who have received copies, or rights, from you under this License will not have their licenses terminated so long as such parties remain in full compliance.

10. FUTURE REVISIONS OF THIS LICENSE

The Free Software Foundation may publish new, revised versions of the GNU Free Documentation License from time to time. Such new versions will be similar in spirit to the present version, but may differ in detail to address new problems or concerns. See http://www.gnu.org/copyleft/. Each version of the License is given a distinguishing version number. If the Document specifies that a particular numbered version of this License "or any later version" applies to it, you have the option of following the terms and conditions either of that specified version or of any later version that has been published (not as a draft) by the Free Software Foundation. If the Document does not specify a version number of this License, you may choose any version ever published (not as a draft) by the Free Software Foundation. ADDENDUM: How to use this License for your documents To use this License in a document you have written, include a copy of the License in the document and put the following copyright and license notices just after the title page: Copyright (c) YEAR YOUR NAME. Permission is granted to copy, distribute and/or modify this document under the terms of the GNU Free Documentation License, Version 1.2 or any later version published by the Free Software Foundation; with no Invariant Sections, no Front-Cover Texts, and no Back-Cover Texts. A copy of the license is included in the section entitled "GNU Free Documentation License". If you have Invariant Sections, Front-Cover Texts and Back-Cover Texts, replace the "with...Texts." line with this: with the Invariant Sections being LIST THEIR TITLES, with the Front-Cover Texts being LIST, and with the Back-Cover Texts being LIST. If you have Invariant Sections without Cover Texts, or some other combination of the three, merge those two alternatives to suit the situation. If your document contains nontrivial examples of program code, we recommend releasing these examples in parallel under your choice of free software license, such as the GNU General Public License, to permit their use in free software.

Printed in Great Britain
by Amazon.co.uk, Ltd.,
Marston Gate.